# An Introduction to Chemistry for Biology Students

## Ninth Edition

# An Introduction to Chemistry for Biology Students

## Ninth Edition

### George I. Sackheim
Associate Professor Emeritus, Chemistry
University of Illinois at Chicago

formerly
Coordinator of Biological and Physical Sciences
Michael Reese Hospital and Medical Center, Chicago

Science Instructor
St. Francis Hospital School of Nursing
Evanston, Illinois

PEARSON

Benjamin
Cummings

San Francisco   Boston   New York
Cape Town   Hong Kong   London   Madrid   Mexico City
Montreal   Munich   Paris   Singapore   Sydney   Tokyo   Toronto

Editorial Director: *Adam Black*
Publisher: *Jim Smith*
Editorial Manager: *Laura Kenney*
Project Editor: *Kristin Rose*
Senior Marketing Manager: *Scott Dustan*
Managing Editor: *Corinne Benson*
Production Supervisor: *Shannon Tozier*
Production Service and Composition: *Nesbitt Graphics, Inc.*
Illustrations: *Nesbitt Graphics, Inc.*
Text Design: *Patrick Devine*
Cover Design: *Seventeenth Street Studios*
Manufacturing Manager: *Pam Augspurger*
Cover Printer: *Phoenix Color*
Printer and Binder: *R. R. Donnelley, Crawfordsville*
Cover Image: *Nick Rowe/Getty Images*
Cover Photo: *Microplate filled with colorful chemicals*

ISBN: 10-digit 0-8053-9571-7
       13-digit 978-0-8053-9571-6

**PEARSON**

**Benjamin Cummings**

4 5 6 7 8 9 10 – DOC – 11 10
www.aw-bc.com

# Contents

# To the Student

*An Introduction to Chemistry for Biology Students*, Ninth Edition, is not an ordinary book. It has been programmed to help you review the basic facts, concepts, and terminology of chemistry that are essential to an understanding of biological phenomena. Today's biology courses place increasing emphasis on the chemical processes that underlie the critical biological functions. This book will help you to understand those processes.

The topics involved are among the most critical and exciting that science will explore in the years ahead. What are the basic chemical processes underlying biological phenomena? What are the essential differences between living and nonliving matter? What are the conditions under which molecules organize into living matter? Can these conditions be duplicated experimentally?

If you have already had a course in chemistry, this programmed book can serve as an effective review of the fundamental concepts. If you have had no previous chemistry, the program will give you the background you need to gain a clear understanding of the biological processes you will be studying.

The material covered in this book will help you most if you complete it during the first two weeks of your biology course. Having done that, you will be ready to handle the chemical aspects of biology as they come up.

When you have completed the program, you may want to repeat certain material. To simplify this process, use the book's Index to help you locate specific topics.

If you follow the directions and complete this program, you will learn to

- recognize the elements present in various compounds

- draw Lewis dot structures

- know what is meant by pH and by ionization

- recognize acids, bases, and salts

- discriminate between electrolytes and nonelectrolytes

- understand osmosis and diffusion

- distinguish between passive and active transport

- understand osmotic pressure

- understand transmission of nerve impulses, including depolarization and repolarization

- understand how neurotransmitters work

- know how DNA replicates

- know how mRNA is formed and how it regulates protein synthesis

- understand oxidation and reduction

- know what isotopes are

- recognize various organic functional groups

- distinguish between types of isomers

- differentiate among carbohydrates, fats, and proteins

- understand how enzymes function

- recognize nucleic acids

- recognize and name methyl compounds

- understand the function of methyltransferases, demethylases, and demethylation

- be able to recognize and name N-methyl compounds

- understand biologic oxidation, including glycolysis, the citric acid cycle, and the electron transport chain

- understand reactions involved in photosynthesis, including light reactions and the Calvin cycle

- follow the flow of oxygen from the lungs to cells and of carbon dioxide from cells to the lungs on gas partial pressure gradients

# How to Use This Book

This type of instructional book may be new to you. Its subject matter has been presented as a series of numbered problems. Each builds on information learned in the preceding problems. For that reason, it is important not to skip around. The sequence of the problems is important because it is programmed to help you learn more efficiently.

## Respond to Every Problem

Some problems present new information; others review material presented earlier. Every problem presents a learning task that requires some response from you.

You may be asked to make any of the following types of responses:

 ❦ write an answer in a blank space

 ❦ label a diagram

 ❦ draw a simple diagram

 ❦ select the correct answer from among several alternatives

 ❦ write a sentence in answer to a question

When you have written or marked your answer, you will want to find out whether you are correct. Programmed instruction provides you with important feedback by giving you easy access to the answers, which are located in shaded areas on the outer side of each page. *Do not look at the correct answer until after you have marked your own answer.* If you look before answering, you will only impair your own learning process.

## Use an Answer Mask

Bound into the back of the book is a sheet of heavy paper that is perforated. Tear off the outer portion of the sheet for use as an answer mask. Here is what to do:

 1. As you start working on the first page, a right-hand page, cover the shaded answer column with the answer mask *before* looking at the

problems. When you turn the page, shift your answer mask to cover the answer column on the left-hand page before reading the problems on that page. *Be sure you have covered the answers before you read anything.*

2. Each problem number is centered on the page with a shaded rectangle behind it. Read the problem carefully, then record your answer. Make sure you either write each answer or do whatever the directions say. Do not simply think of the answer and then go on. Actually writing or marking your answer reinforces your learning.

3. Now move the answer mask aside to reveal the answer, which you will find aligned with the problem number. (If the main part of the problem runs onto the next page, you will find the answer at the top of that page.)

4. If your answer was correct, move on to the next problem.

5. If your answer was incorrect, reread the problem (if necessary, reread several of the preceding problems) until you understand your error and know why the given answer is correct. Then go on.

When you have completely worked your way through this book, you should have the knowledge of chemistry you need to succeed in your biology courses.

To find a topic or to review a topic already completed, look in the index to locate the reference(s) to that subject.

PART I

Inorganic
Chemistry

# Atomic Structure 1

## ELEMENTARY PARTICLES

Atoms are made up of several components. Collectively these components are called the *elementary particles*. We will be discussing the three major elementary particles: *protons, neutrons*, and *electrons*.

Here is a diagram of an atom:

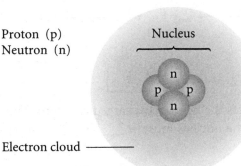

Proton (p)
Neutron (n)

Nucleus

Electron cloud

1

The protons (p) and the neutrons (n) are packed together in an inner core called the _____. The outer part of the atom, which contains electrons, is called the _____.

a negative electrical charge, because the electron cloud consists of electrons

The electron cloud has a negative electrical charge. What type of charge would you expect the electron to have? _____

_____

a. repel
b. attract
c. repel
d. attract

The electron has a *negative electrical charge* and is symbolized by e⁻. Remember that *like* electrical charges repel each other and *unlike* charges attract.

Indicate whether the following pairs of charges would attract or repel each other.

a.   ⊕ ⊕ _____          b.   ⊖ ⊕ _____

c.   ⊖ ⊖ _____          d.   ⊕ ⊖ _____

positive

The nucleus attracts the negatively charged electrons. Therefore, the overall charge of the nucleus must be _____ (negative/positive).

proton

5

The neutron was named for its electrical characteristics. It has no electrical charge; it is neutral. In other words, the positive charge of the nucleus must be due to the *second type of particle* it contains. This second type of particle is the _____.

So far, then, we have this picture of atomic structure:

a. An atom consists of an inner part, or _____, that is made up of _____ and _____.

b. The electron has what type of charge? _____

c. The proton has what type of charge? _____

d. The neutron has a charge of _____.

a. nucleus; protons; neutrons
b. negative
c. positive
d. zero (0)

The charge on the electron balances the charge on the proton. If the electron has a charge of $-1$, then the proton would have a charge of _____ ($-1, +1, \pm1$).

+1

An atom with one proton in its nucleus and one electron outside that nucleus would therefore have an overall charge of _____ ($+1, -1, 0$).

0

**9**

Atoms are electrically neutral. Therefore, an atom will contain: (check one)

___ more protons than electrons
___ more electrons than protons
___ an equal number of protons and electrons

an equal number of protons and electrons

12

a. proton
b. electron

2; 2; 2

An atom with 12 protons in the nucleus would have how many electrons outside the nucleus? _____

The atom with the simplest atomic structure is hydrogen. For simplicity we shall merely indicate the electron(s) outside the nucleus and omit the electron cloud.

$$\text{p} \qquad \text{e}^-$$

Hydrogen

a.  The nucleus of the hydrogen atom consists of one _____.

b.  The outer part of the atom, the electron cloud, contains one

_____.

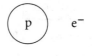

The helium atom is a little more complicated.

$$\text{e}^- \qquad \text{p} \quad \overset{\text{n}}{\underset{\text{n}}{\text{p}}} \qquad \text{e}^-$$

Helium

It contains: (how many?)

_____ neutrons
_____ protons
_____ electrons

# ATOMIC NUMBER

There are over 100 known elements. Each element has two numbers associated with it, numbers that give certain facts about the structure of its atoms.

The first number is the *atomic number*. The atomic number is the number of protons in the nucleus of the atom.

Hydrogen, the simplest atom, contains only one proton, so the atomic number of hydrogen is _____.

1

Uranium is the most complicated of the elements that occur naturally. A uranium atom contains 92 protons, 146 neutrons, and 92 electrons. The atomic number of uranium is _____.

92

---

An atom of magnesium, atomic number 12, must have a nucleus containing _____ protons.

If the nucleus contains 12 protons, then there must be how many electrons? _____

12; 12

Therefore, the atomic number of an element indicates the number of _____ in the nucleus of the atom and also the number of _____ outside the nucleus.

protons; electrons

# MASS NUMBER*

The second number associated with each atom is the *mass number*. The mass number expresses the sum of the masses of the particles in the atom.

A proton has a mass of 1 dalton. An electron is considered to have zero mass, or a mass of 0.

A hydrogen atom has a mass of _____. (If you don't know, see problem 11.)

The helium atom has a mass number of 4.

a.  The 2 protons in the helium atom have a total of how many daltons? _____

b.  The 2 electrons in the helium atom have a total of how many daltons? _____

c.  Therefore, for the helium atom to have a mass number of 4, the 2 neutrons must contain how many daltons? _____

d.  If 2 neutrons have a total of 2 daltons, a neutron must have an atomic mass of _____.

---

* The *atomic mass* of an atom is the weighted average of the masses of its isotopes (see the later section on isotopes). The atomic mass of an atom is a decimal value, as shown in the periodic table on the inside front cover of this book. The *mass number* of an atom is a whole number and is equal to the sum of protons and neutrons in that atom.

Because the electrons, which have practically no mass, are located outside the nucleus, the entire mass of the atom can be considered to be located:

__ in its electron cloud      __ in the nucleus

The *atomic number* indicates the number of protons (each with atomic mass 1) inside the nucleus of an atom. The *mass number* indicates the number of protons and neutrons (each with atomic mass 1) in the nucleus. Therefore, the number of neutrons can be determined by *subtracting* the atomic number from the mass number.

The sodium atom has a mass number of 23 and an atomic number of 11. The number of neutrons in the nucleus of the sodium atom is

_____.

12

The carbon atom has an atomic number of 6 and a mass number of 12. The carbon atom contains: (how many?)

_____ protons in its nucleus
_____ neutrons in its nucleus
_____ electrons outside its nucleus

6; 6; 6

The element phosphorus has an atomic number of 15 and a mass number of 31. Indicate on the blank lines in the diagram the number of protons, neutrons, and electrons.

$\overset{\text{— p}}{\underset{\text{— n}}{\bigcirc}}$ — e⁻

Phosphorus

15p
16n    15e⁻

Diagram the structure of the magnesium atom, atomic number 12 and mass number 24.

Magnesium

---

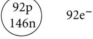

The uranium atom has the atomic structure shown here.

92p
146n          92e$^-$

Uranium

Therefore, the uranium atom has an atomic number of _____ and a mass number of _____.

## ISOTOPES

a.  Draw the structure of a carbon atom, atomic number 6 and mass number 12.

---

12p
12n          12e$^-$

92; 238

a.   6p
     6n          6e$^-$

Carbon-12

b.  Draw the structure of a carbon atom, atomic number 6 and mass
    number 13.

 **26**

Here are the structures you drew for the two carbon atoms:

Carbon-12              Carbon-13

a.  These atoms have _____ (the same/different) atomic
    number(s).

b.  These atoms have _____ (the same/different) mass
    number(s).

Such atoms are called *isotopes.*

 **27**

Isotopes, then, may be defined as atoms that have:

__  the same atomic number and the same mass number
__  different atomic numbers
__  different mass numbers and the same atomic number

b.

6p
7n        6e⁻

Carbon-13

a.  the same
b.  different

different mass numbers
and the same atomic
number

**a.**  17p 18n   17e⁻

Chlorine-35

**b.**  17p 20n   17e⁻

Chlorine-37

**a.**  1p   1e⁻

Hydrogen-1

**b.**  1p 1n   1e⁻

Hydrogen-2

**c.**  1p 2n   1e⁻

Hydrogen-3

**a. same**
**b. different**

## 28

Draw the two isotopes of chlorine, atomic number 17 and mass numbers 35 and 37.

a.  chlorine-35      b.  chlorine-37

## 29

Draw the three isotopes of hydrogen, atomic number 1 and mass numbers 1, 2, and 3.

a.  hydrogen-1      b.  hydrogen-2      c.  hydrogen-3

## 30

*Radioisotopes*, isotopes that give off radiation, are frequently used in medical applications. I-131 (radioactive iodine, also written as $^{131}$I) is used in the diagnosis and treatment of thyroid conditions. How does I-131, radioactive iodine, compare with I-127, nonradioactive iodine, in:

a.  atomic number _____

b.  mass number _____

# ELECTRON ENERGY LEVELS

The electrons are located outside of the nucleus of the atom. These electrons make up the electron cloud, which may be subdivided into different energy levels. The first energy level is nearest the nucleus; then comes the second energy level, the third energy level, and so on.

Each energy level can hold a certain maximum number of electrons. This maximum number may be determined by using the formula $X = 2n^2$ ($X$ is the maximum number of electrons in energy level number $n$).

Using the formula $X = 2n^2$, if $n = 1$, then $X =$ _____. The energy level indicated by $n = 1$ is the first energy level. Therefore, the first energy level can hold a maximum of _____ electrons.

2; 2

a.  Can the first energy level hold fewer than two electrons? _____

b.  Can the first energy level hold more than two electrons? _____

a. yes
b. no

For the second energy level, where $n = 2$, the maximum number of electrons is _____.

8

34

a.  The maximum number of electrons in the third energy level is _____.

b.  The maximum number of electrons in the fourth energy level is _____.

a. 18
b. 32

**2e⁻ in the first energy level; 8e⁻ in the second energy level; 18e⁻ in the third energy level; 32e⁻ in the fourth energy level**

8

Label the maximum number of electrons possible in each energy level in the diagram.

The following rules must be observed when considering the placement of electrons in the various energy levels. The first energy level must be filled with 2 electrons before electrons can go into the second energy level.

The second energy level must be filled with _____ electrons before electrons can go into the third energy level.

For elements having more than three energy levels, the sequence of filling those levels is complex and can be found in a general chemistry text.

On the diagram, show the structure of the hydrogen atom, atomic number 1 and mass number 1.

Hydrogen atom

a.   The number of protons in the hydrogen atom is _____.

b.   The number of neutrons is _____.

c.   The number of electrons is _____.

d.   The 1 electron in the hydrogen atom must go into which energy level: first, second, or third? _____

In the space provided, draw the structure of the helium atom, atomic number 2 and mass number 4.

Helium atom

Hydrogen atom

a.  1
b.  0
c.  1
d.  first

Helium atom

a. 3
b. 4
c. 3

2

Lithium atom

Now let's look at the structure for the lithium atom, atomic number 3 and mass number 7.

a.   The number of protons in the lithium atom is _____.

b.   The number of neutrons in the lithium atom is _____.

c.   The number of electrons in the lithium atom is _____.

40

There are 3 electrons in the lithium atom. How many energy levels will the lithium atom have? _____ (If you aren't sure, check problem 36.)

41

Complete the diagram of the lithium atom.

Lithium atom

## 42

The carbon atom has atomic number 6 and mass number 12.

a.   The carbon atom contains _____ protons.

b.   The carbon atom contains _____ neutrons.

c.   The carbon atom contains _____ electrons.

d.   Complete the structure of the carbon atom.

Carbon atom

**a. 6**
**b. 6**
**c. 6**
**d.**

Carbon atom

## 43

a.   In the element sodium, atomic number 11, there are how many electrons? _____

b.   How many energy levels will the sodium atom have? _____

c.   How many electrons will each energy level of the sodium atom have?

1st energy level ___      2nd energy level ___      3rd energy level ___

d.   Diagram the complete structure of the sodium atom, atomic number 11 and mass number 23.

$$\begin{array}{c}\underline{\phantom{-}} \text{p} \\ \underline{\phantom{-}} \text{n}\end{array}$$

Sodium atom

**a. 11**
**b. 3**
**c. 2; 8; 1**

Sodium atom

Silicon atom

Diagram the structure of the silicon atom, atomic number 14 and mass number 28.

# Chemical Symbols 2

Elements are the building blocks of the universe. They cannot be broken down into simpler substances by ordinary chemical means. Each of the over 100 elements has its own name and its own properties. The symbol for an element is usually an abbreviation for its name. Frequently the abbreviation is merely the first letter of that name. The following table lists some of these elements.

| Element | Symbol |
| --- | --- |
| Hydrogen | H |
| Carbon | C |
| Oxygen | O |
| Nitrogen | N |
| Phosphorus | P |
| Sulfur | S |
| Iodine | I |

What elements are present in water, $H_2O$? _____

hydrogen and oxygen

What elements are present in ammonia, $NH_3$? _____

nitrogen and hydrogen

**carbon, hydrogen, and oxygen**

What elements are present in glucose, $C_6H_{12}O_6$? _____

**carbon, hydrogen, oxygen, nitrogen, and sulfur**

Biotin, one of the B-vitamins, $C_{10}H_{16}O_3N_2S$, contains which elements?

_____

**carbon, hydrogen, oxygen, nitrogen, and phosphorus**

Adenosine triphosphate, ATP, the body's primary energy compound, has the formula $C_{11}H_{18}O_{13}N_5P_3$. What elements does it contain? _____

**Ca; Br; Si; Ba**

When the names of more than one element begin with the same letter, frequently the second letter of the name is added to the symbol. Note that only the first letter of the symbol is capitalized.

Here are some of the elements whose symbols are the first two letters of their names. Write the symbols for each.

| Element | Symbol |
|---------|--------|
| Calcium | _____ |
| Bromine | _____ |
| Silicon | _____ |
| Barium | _____ |

**calcium and bromine**

51

$CaBr_2$ contains which two elements? _____

BaO$_2$ contains which two elements? _____

barium and oxygen

CaC$_2$O$_4$ contains which three elements? _____

calcium, carbon, and oxygen

There are several elements whose symbols are not derived from the first letter or first two letters of their English names. Some of those elements are listed in the following table.

| Element | Symbol |
|---------|--------|
| Chlorine | Cl |
| Sodium | Na (from the Latin *natrium*) |
| Magnesium | Mg |
| Potassium | K (from the Latin *kalium*) |
| Zinc | Zn |
| Iron | Fe (from the Latin *ferrum*) |

54

Write the name of the element represented by each of the following symbols:

H _____     Na _____     Si _____
N _____     K _____     Br _____
O _____     C _____     Zn _____
Ca _____     S _____     Fe _____

*Hydrogen; Nitrogen; Oxygen; Calcium; Sodium; Potassium; Carbon; Sulfur; Silicon; Bromine; Zinc; Iron*

a. potassium and
   bromine
b. carbon and chlorine
c. sodium, oxygen, and
   hydrogen
d. magnesium and
   chlorine
e. sodium, nitrogen,
   and oxygen
f. potassium, iron,
   carbon, and nitrogen

a. magnesium and
   nitrogen

What elements are present in each of the following?

a.   KBr _____

b.   $CCl_4$ _____

c.   NaOH _____

d.   $MgCl_2$ _____

e.   $NaNO_3$ _____

f.   $K_3FeC_6N_6$ _____

**56**

In addition to carbon, hydrogen, and oxygen, which elements are present in the following compounds whose structures are indicated?

a.   chlorophyll a _____

b.   heme (a part of the hemoglobin molecule) _____          **b.  iron and nitrogen**

c.   T$_4$, a thyroid hormone _____                               **c.  iodine and nitrogen**

d.   cysteine, a nonessential aminoacid _____                     **d.  sulfur and nitrogen**

# Atoms and Molecules 3

### 57

One atom of hydrogen is written as H. The number 1 is understood. Two atoms of hydrogen are written as 2H. Three atoms of carbon are written as _____.

3C

### 58

Atoms combine to form *molecules*. A molecule contains two or more atoms of the same or different elements. The molecule $O_2$ contains how many atoms of oxygen? _____

2

### 59

Each molecule of glucose, $C_6H_{12}O_6$, contains:

_____ atoms of carbon
_____ atoms of hydrogen
_____ atoms of oxygen

6; 12; 6

### 60

A water molecule is composed of two hydrogen atoms and one oxygen atom. Write the formula for a water molecule. _____

$H_2O$

$C_{12}H_{22}O_{11}$

$C_{63}H_{90}O_{14}N_{14}PCo$

a. 6
b. $8H_2O$

molecule; hydrogen; 2

### 61

A molecule of sucrose contains 12 atoms of carbon, 22 atoms of hydrogen, and 11 atoms of oxygen. The formula for sucrose is _____.

### 62

One molecule of vitamin $B_{12}$ contains:

63 atoms of carbon
90 atoms of hydrogen
14 atoms of oxygen
14 atoms of nitrogen
 1 atom of phosphorus
 1 atom of cobalt (Co)

The formula for vitamin $B_{12}$ is _____.

### 63

When you want to represent more than one molecule of a compound, the number of molecules precedes the symbols.

a.   Thus, $6CO_2$ indicates how many molecules of carbon dioxide?

_____

b.   How would eight molecules of water be represented? _____

### 64

The formula $H_2$ indicates one _____ of _____, composed of how many atoms? _____

65

The formula $2H_2$ indicates _____ of hydrogen.

two molecules

66

What do the following formulas represent?

3O _____     $4N_2$ _____     $7H_2O$ _____

3 atoms of oxygen;
4 molecules of nitrogen;
7 molecules of water

# Ionization $4$

## IONS

**67**

Atoms contain equal numbers of protons and electrons. Thus, atoms are:

__ electropositive     __ electronegative     __ electrically neutral

**68**

The electron is a _____ (negatively/positively) charged particle. So, if an atom loses an electron, it will then have an overall _____ (negative/positive) charge.

**69**

If an atom were to *gain* an electron, it would then have an overall _____ charge.

**70**

An *ion* is an atom that has acquired an electrical charge by either losing or gaining a(n) _____.

electrically neutral

negatively; positive

negative

electron

**positively**

Note that electrons, which are located outside the nucleus, are lost or gained in the formation of ions. The nucleus, which contains protons and neutrons, does not take part in the formation of ions.

If a hydrogen atom loses its electron, it becomes a hydrogen ion. Because it loses an electron, it is a _____ (negatively/positively) charged ion.

**a. ion**
**b. negatively**
**c. positively**

a.    Any atom that has lost or gained an electron is a(n) _____.

b.    If an atom gains an electron, it becomes a _____ charged ion.

c.    If an atom loses an electron, it becomes a _____ charged ion.

**valence shell**

Some elements reach their most stable state when they have 2 electrons in their outer (their only) energy level, which is called the *valence shell*.

$$2p$$
$$2n$$
$$2e^-$$

Helium atom

This atom of helium, for example, contains 2 electrons, which fill the first energy level. So, the first energy level in helium is called the

_____.

Helium (He) has a completed valence shell, one that can hold no more electrons. Thus, helium is a very _____ (stable/unstable) atom. Such elements are usually unreactive.

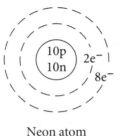

75

Most atoms reach their most stable state when they have 8 electrons in their outer energy level. This property is called the octet rule.

Neon atom

This atom of neon (Ne), for example, contains 10 electrons. Two of the electrons fill the first energy level, and the remaining 8 electrons are in the second energy level. So the second energy level of neon would be called its _____.

**stable**

**valence shell**

# LEWIS DOT STRUCTURE

### 76

He :

Atomic structures such as that of neon above may be abbreviated with the symbol representing the nucleus and all of the inner electrons. Each outer electron, those in the valence shell, is indicated by a dot.

Such a representation is called a *Lewis dot structure.*

The electron-dot structure for neon, which has 8 electrons in its valance shell, is

$$: \overset{..}{\underset{..}{Ne}} :$$

The Lewis dot structure for helium is _____ (see problem 73).

### 77

**a.** valence shell; third; 1
**b.** Cl
**c.**
$$: \overset{..}{\underset{..}{Cl}} \cdot$$

Look at this atom of chlorine.

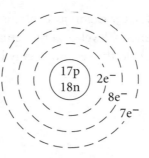

Chlorine atom

a.  In chlorine, the outer energy level, or _____, is the _____ energy level. It contains 7 electrons. Thus, to attain a stable state, the chlorine atom needs to gain how many electrons? _____

b.  What is the symbol for chlorine? _____

c.  What is the Lewis dot structure for chlorine? _____

**78**

Suppose that an atom of chlorine gains the 1 electron it needs to become stable.

a.   Because it has gained an electron, it is now a charged atom, called a(n) _____.

b.   Would it have a positive or negative charge? _____

a. ion
b. negative

**79**

Which of the following symbols stands for the chloride ion (a chlorine atom that has gained 1 electron)?

__ Cl        __ $Cl^+$        __ $Cl^-$

$Cl^-$

**80**

If a chlorine atom is to gain an electron, some other atom must lose the electron.

a.   That atom could be sodium, the symbol for which is _____.

b.   The sodium atom has how many electrons in its valence shell?

_____

a. Na
b. 1
c. Na ·
d. lose

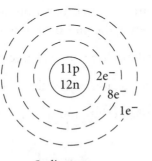

Sodium atom

c.   The Lewis dot structure for sodium is _____.

d.   To become stable, it will have to _____ (gain/lose) 1 electron.

**positively; ion**

**Na⁺**

When the sodium atom loses 1 electron, it becomes a _____ (positively/negatively) charged _____.

Which symbol should we use for the sodium ion?

__ Na    __ Na⁺    __ Na⁻

The reactions for the gain of an electron by the chlorine atoms may be written as

$$:\overset{..}{\underset{..}{Cl}}\cdot + 1e \longrightarrow :\overset{..}{\underset{..}{Cl}}:^{-}$$

which can be simplified to

$$:\overset{..}{\underset{..}{Cl}}\cdot + 1e \longrightarrow Cl^{-}$$

where the dots are assumed and not written.

**a. +1**
**b. −1**
**c. +2**

Positive and negative charges are indicated by superscript plus (positive) or minus (negative) signs, respectively, with the number 1 being understood. Charges greater than 1 are written as superscripts, with the number preceding the plus or minus sign.

a.  $Na^+$ indicates a sodium ion with a charge of _____.

b.  $Cl^-$ indicates a chloride ion with a charge of _____.

c.  $Mg^{2+}$ indicates a magnesium ion with a charge of _____.

**a. Al³⁺**
**b. S²⁻**

a.  How would an aluminum ion with a charge of $+3$ be indicated?

    _____

b.  How would a sulfide ion with a charge of $-2$ be indicated?

    _____

$$Na + Cl \longrightarrow Na^+ + Cl^-$$

Reaction of sodium and chlorine

When a sodium atom reacts with a chlorine atom, the sodium atom
_____ (gains/loses) 1 electron to form a sodium ion with a
_____ charge of _____ (how many?).

At the same time, the chlorine atom _____ (gains/loses)
1 electron to form an ion with a _____ charge of _____
(how many?).

This reaction may be written as

$$Na \cdot + \cdot \overset{\cdot\cdot}{\underset{\cdot\cdot}{Cl}} : \longrightarrow Na^+ + Cl^-$$

**loses; positive; 1;
gains; negative; 1**

## 86

Look at this magnesium atom. Its symbol is _____.

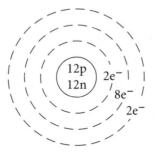

Magnesium atom

It has _____ electrons in its valence shell. So, to gain
stability, it will have to _____ (gain/lose) _____
(how many?) electrons.

The Lewis dot structure for magnesium is _____.

The symbol for the magnesium ion is _____ ($Mg^{2-}$, Mg,
$Mg^{2+}$).

**Mg; 2; lose; 2; Mg: ;
$Mg^{2+}$**

−1; +2; $Mg^{2+}+Cl^-+Cl^-$
or $Mg^{2+}+2Cl^-$

The reaction between magnesium and chlorine may be written as

$$Mg + Cl_2 \longrightarrow Mg^{2+} + 2Cl^-$$

This reaction shows that 1 chlorine molecule reacts with 1 magnesium atom to form 2 chloride ions, each with a charge of _____, and 1 magnesium ion with a charge of _____.

This reaction may be written as

a.   The sulfur atom, atomic number 16, has how many energy levels? _____

a.  3
b.  6
c.  −2
d.  $\overset{\cdot\cdot}{S}\!:$

e.  $S^{2-}$

b.   How many electrons are in its valence shell? _____

c.   If a sulfur atom gains 2 electrons to fill its valence shell, then it will have a charge of _____.

d.   What is the Lewis dot structure for sulfur? _____

e.   The symbol for the sulfide ion is _____.

a.  $:\!\overset{\cdot\cdot}{\underset{\cdot\cdot}{F}}\!\cdot$   b.  $\cdot\overset{\cdot}{\underset{\cdot}{C}}\cdot$

Write the Lewis dot structure for each of the following.

a.   fluorine, atomic number 9 _____

b.   carbon, atomic number 6 _____

## 90

Complete the Lewis dot structure for the following reaction:

$$\text{Li} \cdot + \cdot \overset{\cdot\cdot}{\underset{\cdot\cdot}{\text{Br}}} : \longrightarrow$$

## 91

How can we tell whether an atom will lose or gain electrons to reach a stable structure of 8 electrons in its valence shell? Look over the following diagrams.

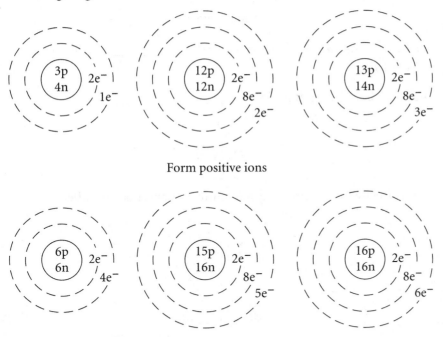

Form positive ions

Do not form positive ions

In general, an atom with _____ (how many?) electrons in its valence shell tends to lose electrons and form a _____ (positively/negatively) charged ion.

**1; loss; Na$^+$**

**92**

Sodium has atomic number 11. How many electrons does it have in its valence shell? _____

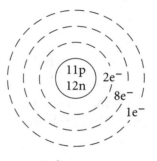

Sodium atom

The sodium ion therefore, is formed by the _____ (loss/gain) of 1 electron.

The symbol for the sodium ion is _____.

**+1; H$^+$**

**93**

Here is the hydrogen atom, atomic number 1 and mass number 1.

Hydrogen atom

Hydrogen usually loses its 1 electron to form a hydrogen ion with a charge of _____.

The hydrogen ion can be represented as _____.

Look at the following diagrams carefully.

Form negative ions

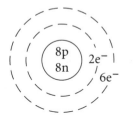

Do not form negative ions

In general, atoms that have _____ (how many?) electrons in their valence shells form negatively charged ions. When an atom gains electrons to fill its valence shell, it forms _____ charged ions.

**6 or 7; negatively**

9; 7; gain; 1; F⁻

Complete the following for the fluorine (F) atom shown here:

Fluorine atom

Atomic number: _____
Number of electrons in valence shell: _____
Ion formed by _____ (gain/loss) of _____ electron(s)
Symbol for ion: _____

S; 16; 32; S²⁻

Complete the following for the sulfur atom shown here:

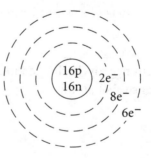

Sulfur atom

Symbol for sulfur: _____
Atomic number: _____
Mass number: _____
Symbol for ion: _____

Atoms with 4 or 5 electrons in their valence shells usually do not form ions. We shall discuss these ions later.

Atoms with 8 electrons in their valence shells are stable. Because they do not have to gain or lose electrons, you would expect these stable atoms to be

__ reactive       __ inert (unreactive)

**inert (unreactive)**

The hydrogen atom, atomic number 1, has how many electrons in its valence shell? _____

The hydrogen atom will tend to lose the 1 electron in its valence shell to form an ion with a charge of _____.

1; +1

99

The chlorine atom, which has the symbol _____, has atomic number 17. It has how many electrons in its valence shell? _____

The chlorine atom will tend to gain _____ (how many?) electron(s) to form an ion with a charge of _____.

Cl; 7; 1; −1

# IONIC BONDS

**gains**

### 100

Consider the reaction between sodium and chlorine:

$$Na + Cl \longrightarrow Na^+ + Cl^-$$

The sodium ion and the chloride ion are oppositely charged. These ions are held together by the attraction of their opposite charges. We say that there is an *ionic bond* between the sodium ion and the chloride ion.

An ionic bond is produced whenever one atom loses an electron or electrons and another atom _____ an electron or electrons.

**ionic bond**

### 101

In the reaction $Zn + S \longrightarrow Zn^{2+} + S^{2-}$, the zinc ion and the sulfur ion are held together by a (n) _____.

**ion; negatively; ion**

### 102

When a sodium atom combines with a chlorine atom according to the equation $Na + Cl \longrightarrow Na^+ + Cl^-$, a compound containing a positively charged sodium _____ and a _____ charged chloride _____ is formed. This compound (sodium chloride) is usually written as NaCl, with the charges being understood and not written.

**chloride; $Cl^-$**

### 103

When the compound NaCl is placed in water, the ionic bond holding the sodium ion and the chloride ion together is weakened, so that these ions are free to move throughout the solution.

Therefore, NaCl in water produces a *solution* containing sodium ions ($Na^+$) and _____ ions (_____).

# ACIDS, BASES, AND SALTS

## 104

a. When the compound HCl is placed in water, it produces a solution containing _____ ions (_____) and _____ ions (_____).

b. When the compound $H_2SO_4$ is placed in water, it produces a solution containing _____ ions and $HSO_4^-$ ions.

a. **hydrogen; $H^+$; chloride; $Cl^-$**
b. **hydrogen ($H^+$)**

## 105

Any substance that yields hydrogen ions ($H^+$) in solution is called an *acid*.

$$HCl \longrightarrow H^+ + Cl^-$$
$$H_2SO_4 \longrightarrow H^+ + HSO_4^-$$
$$HNO_3 \longrightarrow H^+ + NO_3^-$$

$HNO_3$ is an *acid* because it yields _____ in solution.

**hydrogen ions or $H^+$**

## 106

A hydrogen ion is also a *proton*. Therefore, acids yield hydrogen ions or _____ in solution.

**protons**

acids

## 107

Note that hydrogen ions associate with water molecules to form a "hydronium" ion $(H_3O)^+$. Nonetheless, we will use the term *hydrogen ion* $(H^+)$ in this book.

Substances that yield hydrogen ions or protons in solution are called _____. Substances that accept (react with) hydrogen ions or protons are called *bases*.

bases

## 108

When sodium hydroxide, NaOH, is placed in water, sodium ions and hydroxide ions are present:

$$NaOH \longrightarrow Na^+ + OH^-$$

The hydroxide ions $(OH^-)$ react with hydrogen ions to form water:

$$OH^- + H^+ \longrightarrow H_2O$$

Because hydroxide ions accept (react with) hydrogen ions, $OH^-$ ions are _____.

yes

## 109

Potassium hydroxide, KOH, also yields $OH^-$ ions in water. Would KOH be a base? _____

bases; they accept (react with) hydrogen ions

## 110

Bicarbonate ions, $HCO_3^-$, are important ions in body fluids. They react as follows:

$$HCO_3^- + H^+ \longrightarrow H_2CO_3$$

Are bicarbonate ions acids or bases? _____

Why? _____

### 111

Ammonia is a waste product of the body's metabolism of protein.
Ammonia reacts as follows:

$$NH_3 + H^+ \longrightarrow NH_4^+$$

Ammonia                    Ammonium ion

Is ammonia an acid or a base? _____

base

### 112

Acids _____ hydrogen ions or protons.

Bases _____ hydrogen ions or protons.

yield; accept (react with)

### 113

When the compound NaCl is placed in water, it yields _____
ions and _____ ions.

sodium (or $Na^+$);
chloride (or $Cl^-$)

### 114

The solution of NaCl in water would not be considered an acid because:

__ it yields $OH^-$ ions
__ it doesn't break up into ions
__ it yields no hydrogen ions

it yields no hydrogen
ions

a. salt
b. acid
c. salt
d. acid
e. base
f. acid
g. base

A compound that yields ions other than hydrogen ions ($H^+$) or hydroxide ions ($OH^-$) is called a *salt*. Are the following underlined substances acids, bases, or salts?

a.  $\underline{Na_2CO_3}$  $\longrightarrow$  $2Na^+ + CO_3^{2-}$  _____

b.  $\underline{H_2SO_4}$  $\longrightarrow$  $H^+ + HSO_4^-$  _____

c.  $\underline{MgCl_2}$  $\longrightarrow$  $Mg^{2+} + 2Cl^-$  _____

d.  $\underline{H_2PO_4^-}$  $\longrightarrow$  $H^+ + HPO_4^{2-}$  _____

e.  $\underline{NaOH} + H^+$  $\longrightarrow$  $Na^+ + H_2O$  _____

f.  $\underline{H_3PO_4}$  $\longrightarrow$  $H^+ + H_2PO_4^-$  _____

g.  $\underline{Ca(OH)_2} + 2H^+$  $\longrightarrow$  $Ca^{2+} + 2H_2O$  _____

# ELECTROLYTES
# AND NONELECTROLYTES

yes

Acids, bases, and salts are called *electrolytes*. Solutions of electrolytes conduct electricity because of the presence of ions:

$$HCl \longrightarrow H^+ + Cl^-$$

Hydrochloric
acid

Would a solution of hydrochloric acid be an electrolyte? _____

### 117

Here's what happens when magnesium chloride is put in solution:

$$MgCl_2 \longrightarrow Mg^{2+} + 2Cl^-$$

a.  Does the solution of $MgCl_2$ contain ions? _____

b.  Is $MgCl_2$ an acid, a base, or a salt? _____

c.  Is a solution of $MgCl_2$ an electrolyte? _____

a.  yes
b.  salt
c.  yes

### 118

When sucrose, $C_{12}H_{22}O_{11}$, is placed in water, no ions are produced.

a.  Would a solution of sucrose conduct electricity? _____

b.  Would a solution of sucrose be an electrolyte or a nonelectrolyte?

_____

a.  no
b.  nonelectrolyte

### 119

Which of the following solutions would be electrolytes?_____

a.  $MgSO_4 \longrightarrow Mg^{2+} + SO_4^{2-}$

b.  alcohol $\longrightarrow$ no ions

c.  $KOH \longrightarrow K^+ + OH^-$

d.  $C_6H_{12}O_6$ (glucose) $\longrightarrow$ no ions

e.  $HNO_3 \longrightarrow H^+ + NO_3^-$

a, c, e

# ANIONS AND CATIONS

Consider the following electrolytic system, a battery whose two electrodes are immersed in a solution containing the indicated ions.

a.   The anode will attract ions of what charge? _____

b.   The cathode will attract ions of what charge? _____

a.   –
b.   +

a.   Ions attracted toward an anode are called *anions*. Anions have a _____ charge.

b.   Ions attracted toward a cathode are called *cations*. Cations have a _____ charge.

a.   –
b.   +

Na⁺, K⁺, and H⁺

Which are the cations in the solution indicated in problem 120?

_____

## 123

Which are the anions in that same solution? _____

$Cl^-$, $Br^-$, and $SO_4^{2-}$

## 124

a. The principal cation inside animal cells (the principal *intra*cellular cation), the potassium ion, has a _____ (positive/negative) charge.

b. The principal cation outside animal cells (the principal *extra*cellular cation), the sodium ion, has a _____ (positive/negative) charge.

c. The iodide ion, $I^-$, is necessary for the proper functioning of the thyroid gland. The iodide ion is a(n) _____ (anion/cation).

a. **positive**
b. **positive**
c. **anion**

# pH

Mathematically, $pH = -\log H^+$. A log, or logarithm, is an exponent. The log (exponent) of $10^{-3}$ is $-3$.

## 125

The log of $10^{-7}$ is _____.

$-7$

## 126

If $H^+ = 10^{-4}$, then $pH = -\log H^+ = -(-4) = $ _____.

$4$

a. 12
b. 1

### 127

That is, pH is the negative log of the $H^+$.

a.  If $H^+ = 10^{-12}$, pH = _____.

b.  If $H^+ = 10^{-1}$, pH = _____.

a. acid
b. neutral
c. acid

### 128

The acid or basic strength of a solution may be expressed in terms of a number called the *pH* of that solution. The pH scale expresses the concentration of hydrogen ions (and hydroxide ions) in solution.

The pH range is from 0 to 14, with a pH of 7 indicating a neutral solution. A pH below 7 indicates an *acid* solution.

a.  A pH of 3 indicates a(n) _____ solution.

b.  A pH of 7 indicates a(n) _____ solution.

c.  A pH of 6 indicates a(n) _____ solution.

a. 7
b. neutral

### 129

Water is very slightly ionized: $H_2O \longrightarrow H^+ + OH^-$. The hydrogen ion concentration ($H^+$) in water is $10^{-7}$.

a.  What is the pH of water? _____

b.  Is water acid, basic, or neutral? _____

Although all pH values below 7 indicate acid solutions, there is a definite progression of acid strengths according to pH values.

A pH between 5 and 7 indicates a weak acid solution, between 2 and 5 indicates a moderately strong acid solution, and between 0 and 2 indicates a strong acid solution, as shown in the following chart:

a. Which of the following pH values indicates a solution containing a strong acid?

__ 4  __ 7  __ 1  __ 6

b. Which of the following pH values indicates a solution containing a moderately strong acid?

__ 4  __ 7  __ 1  __ 6

c. A pH of 0 indicates what strength acid solution? _____

d. A pH of 3 indicates what strength acid solution? _____

a. 1
b. 4
c. **strong**
d. **moderately strong**

pH values may be indicated as decimal values as well as whole numbers. Thus, a pH of 2.56 indicates a solution whose pH lies between 2 and 3. Thus, this solution is a:

__ strong acid solution
__ weak acid solution
__ moderately strong acid solution

**moderately strong
acid solution**

**6.27**

**1.72**

**7.00**

## 132

Among the following pH values, which solution contains a weak acid?

___ 1.72      ___ 3.75      ___ 2.00      ___ 6.27

## 133

Among the following pH values, which solution contains a strong acid?

___ 1.72      ___ 3.75      ___ 7.00      ___ 5.00      ___ 6.38

## 134

Of solutions with the following pH values, which one is neutral?

___ 2.70      ___ 4.65      ___ 5.00      ___ 7.00

A solution whose pH is above 7 is called a *basic* solution.

A solution whose pH is between 7 and 9 is called a weak basic solution, between 9 and 12 a moderately strong basic solution, and between 12 and 14 a strong basic solution, as shown in the following chart:

Of the following pH values, which indicates a strong basic solution?

__ 3.5    __ 7.0    __ 9.5    __ 13.9    __ 6.4

### 136

Of the following pH values, which indicates a weak basic solution?

__ 2.7    __ 8.3    __ 4.2    __ 10.8    __ 14.0

### 137

Which of the following pH values indicates a moderately strong basic solution?

__ 2.2    __ 4.7    __ 11.1    __ 13.7

### 138

The pH of saliva lies between 5.5 and 6.9, indicating a _____ solution.

### 139

Bile has a pH range of 7.8 to 8.6, so it is a _____ solution.

**13.9**

**8.3**

**11.1**

**weak acid**

**weak basic**

weak acid

strong acid

weak basic

weak basic

a. 10
b. weak

Urine has a pH range of 5.5 to 6.9, so it is a _____ solution.

The gastric juices have a pH range of 1.6 to 1.8, so they make up a _____ solution.

Blood has a pH range of 7.35 to 7.45, so it is a _____ solution.

### 143

The pancreatic juices have a pH range of 7.5 to 8.0, so they make up a _____ solution.

### 144

A difference of 1 in pH is equivalent to a 10-fold difference in acid or base strength. That is, an acid of pH 2 is 10 times as strong as an acid of pH 3. Likewise, a base of pH 10 is 10 times as strong as a base of pH 9.

a.  A solution of pH 4.73 is _____ times as strong as one of pH 5.73.

b.  A solution of pH 9.81 is 10 times as _____ (strong/weak) as one of pH 10.81.

Thus, a small difference in pH can represent a _____
(considerable/negligible) difference in acid or base strength.

Electrode —
Solution —

## BUFFERS

Recall that acids yield hydrogen ions ($H^+$) or protons in solution.
Both hydrochloric acid (HCl) and acetic acid ($CH_3COOH$) are acids,
as indicated by the following reactions:

$$HCl \longrightarrow H^+ + Cl^-$$
$$CH_3COOH \rightleftharpoons CH_3COO^- + H^+$$

The HCl, however, is 100% ionized, as shown by the one-way arrow,
whereas $CH_3COOH$ is less than 1% ionized, as shown by the double
(equilibrium) arrows. That is, relatively speaking, HCl yields many
$H^+$ ions in solution, whereas $CH_3COOH$ yields few $H^+$ ions.

**considerable**

a. few
b. strong
c. weak

Hydrochloric acid, HCl, is called a *strong acid*. It yields relatively many $H^+$ ions in solution.

a.　Acetic acid, $CH_3COOH$, is called a *weak acid*. It yields relatively _____ (many/few) $H^+$ ions in solution.

b.　Sulfuric acid, $H_2SO_4$, yields relatively many $H^+$ ions in solution. It is a _____ (strong/weak) acid.

c.　Citric acid, which yields relatively few $H^+$ ions in solution, is a _____ (strong/weak) acid.

### 147

A base accepts or reacts with protons (hydrogen ions).

a. many
b. few

a.　A strong base reacts with relatively _____ (many/few) $H^+$ ions.

b.　A weak base reacts with relatively _____ (many/few) $H^+$ ions.

### 148

a. strong
b. weak

a.　NaOH, sodium hydroxide, reacts with relatively many $H^+$ ions. It is a _____ (strong/weak) base.

b.　Sodium bicarbonate, $NaHCO_3$, reacts with relatively few $H^+$ ions. It is a _____ (strong/weak) base.

## 149

a.  When a strong acid is added to water, the pH of the solution should go _____ (up/down).

b.  When a strong base is added to water, the pH of the solution should go _____ (up/down).

The addition of a weak acid or a weak base to water should have very little effect on the pH.

## 150

A *buffer solution* is one that maintains a constant pH upon the addition of small amounts of either acid or base. A buffer consists of a weak acid and a weak base. A buffer solution can "sponge up" excess $H^+$ if added to a solution, or it can release $H^+$ if the $H^+$ concentration drops. One such buffer in the body is the $H_2CO_3/NaHCO_3$ system.

Suppose that a strong acid such as HCl is added to the buffer. The reaction is

$$HCl + NaHCO_3 \longrightarrow H_2CO_3 + NaCl$$

Strong acid    Weak base      Weak acid    Salt

Thus, the strong acid has been changed to a weak acid. Should the pH change? _____

## 151

If a strong base such as sodium hydroxide, NaOH, is added to the buffer, then the reaction is

$$NaOH + H_2CO_3 \longrightarrow NaHCO_3 + H_2O$$

Strong base    Weak acid      Weak base    Water

The strong base has been changed to a _____ base. Should the pH change? _____

**it should not change it**

**no**

**no effect**

**none**

**yes**

Another buffer system in the body is the phosphate buffer system, which consists of $NaH_2PO_4/Na_2HPO_4$. What effect will the addition of a strong acid have on the pH of this buffer? _____

153

Should the addition of base affect the pH of a phosphate buffer system? _____

154

What effect should the removal of acid (which is equivalent in effect to the addition of base) have on the pH of a buffer solution? _____

155

The pH of the blood (a buffer solution) lies in the range of 7.35 to 7.45.

Normal metabolic processes add acid to the bloodstream. What effect should these acids have on the pH of the blood? _____

156

Protein buffers are also present in the body. Should they behave similarly to the bicarbonate and phosphate buffers? _____

# Liquid Mixtures 5

## SOLUTIONS

A solution consists of a *solute* (a relatively less abundant substance) dissolved in a *solvent* (a relatively more abundant substance). In living organisms, the solvent is water.

a.  When oxygen dissolves in water, oxygen is the _____ (solute/solvent).

b.  When alcohol is dissolved in water, alcohol is the _____ (solute/solvent).

c.  When salt is dissolved in water, water is the _____ (solute/solvent).

158

A solution consists of a _____ and a _____.

solution

Solute + solvent ⟶ _____.

a. yes
b. homogeneous
c. transparent

Solutions are transparent and homogeneous. *Homogeneous* means "the same throughout," as compared with *heterogeneous*, which means "not the same throughout."

a.   In a saltwater solution, is one part the same as any other part? _____

b.   Solutions are _____ (homogeneous/heterogeneous).

c.   Solutions are _____ (transparent/opaque).

a. yes
b. yes

**161**

In general, solutions pass through membranes.

a.   If glucose is administered intravenously, will the glucose solution pass through the membranes in the body? _____

b.   Will saline (salt) solution pass through a membrane? _____

water

For medical use, solutions are frequently described in terms of *percent* or *parts per hundred.*

In a percent solution, the percentage number indicates the number of grams of solute present in 100 mL of solution. The solvent is usually _____.

a.  A 5% glucose solution contains 5 g of glucose in _____ mL of solution.

b.  A 2% boric acid solution contains _____ g of boric acid in _____ mL of solution.

Physiologic saline solution, 0.95% NaCl solution, contains _____ g of NaCl in _____.

### 165

Another method of expressing solution concentration is in terms of *molarity*, or *moles* of solute per liter of solution. A *mole* of a compound is the number of grams of that compound equal to its molecular mass.

The atomic mass of hydrogen is 1 and that of oxygen is 16. The molecular mass of water, $H_2O$, which consists of 2H and 1O, is $(2 \times 1) + (1 \times 16)$, or _____. Therefore, one mole of water has a mass of _____ g.

### 166

What is the molecular mass of glucose, $C_6H_{12}O_6$? _____
Atomic masses are C = 12, H = 1, O = 16.

a.  100
b.  2; 100

0.95; 100 mL of solution

18; 18

180 g

a. 1; 1 liter
b. 120; 1 liter
c. 0.1; solution
d. 5.85

### 167

A 1-molar (1M) glucose solution contains 1 mole (180 g) of glucose dissolved in enough water to make 1 liter of solution.

Use the following atomic masses in making calculations:

| | | | |
|---|---|---|---|
| H | 1 | O | 16 |
| Mg | 24 | S | 32 |
| Na | 23 | Cl | 35.5 |

a. To prepare a 1M $MgSO_4$ solution, take _____ mole(s) of $MgSO_4$ and dissolve in enough water to prepare _____ (how much?) of solution.

b. To prepare a 1M $MgSO_4$ solution, take _____ g of $MgSO_4$ and dissolve in enough water to make _____ (how much?) of solution.

c. To prepare a 0.1M NaCl solution, take _____ mole(s) of NaCl and dissolve in enough water to make 1 liter of _____.

d. To prepare a 0.1M NaCl solution, take _____ g of NaCl and dissolve in enough water to make 1 liter of solution.

### 168

*Osmolarity* is a method of expressing solution concentration based on the number of particles in solution.

$$\text{Osmolarity} = \text{molarity} \times \frac{\text{number of particles}}{\text{molecule}}$$

Osmolarity is expressed in the units osmoles, or *osmol.*

Consider the following reactions:

$$NaCl \longrightarrow Na^+ + Cl^-$$

1 molecule $\longrightarrow$ 2 particles

$$Glucose \longrightarrow glucose$$

1 molecule $\longrightarrow$ 1 particle

$$K_2SO_4 \longrightarrow 2K^+ + SO_4^{2-}$$

1 molecule $\longrightarrow$ 3 particles

a.   A 1-molar (1M) glucose solution has a concentration of _____ osmol.

b.   A 0.5M NaCl solution has a concentration of _____ osmol.

c.   A 0.4M $K_2SO_4$ solution has a concentration of _____ osmol.

a.  1
b.  1
c.  1.2

### 169

Osmolarity may also be expressed in the units milliosmoles, or *mosmol*, where

$$1 \text{ osmol} = 1000 \text{ mosmol}$$

To change osmoles to milliosmoles, multiply by 1000.

a.   A 0.15M NaCl solution has a concentration of _____ osmol.

b.   A 0.15M NaCl solution has a concentration of _____ mosmol.

Note that human body fluids have an osmolarity of approximately 300 milliosmol.

a.  0.30
b.  300

# SUSPENSIONS

## 170

When sand is placed in water and shaken, a sand-water suspension is produced.

a.   When the shaking is stopped, does the sand settle? _____

b.   Does sand dissolve in water? _____

c.   In a suspension, is there a solute? _____ Why or why not?

_____

Note that particles in a suspension are so large that they settle under the influence of gravity.

## 171

a.   A suspension is _____ (homogeneous/heterogeneous).

b.   A solution is _____ (homogeneous/heterogeneous).

c.   Is a suspension transparent? _____

d.   Is a solution transparent? _____

## 172

Blood is a suspension of blood cells in blood plasma. If blood is treated so that it doesn't clot and is then placed in a test tube, will the blood cells settle? _____

What causes the blood cells to settle? _____

# COLLOIDS
# (COLLOIDAL DISPERSIONS)

Colloids consist of very tiny particles suspended in a liquid, usually water. (Colloids may also be suspended in solids and in gases.) Colloids differ from suspensions in that colloids do not settle. Colloidal dispersions are usually translucent or milky white. They do not pass through membranes.

173

One example of a colloid is protein.

a.   A colloidal dispersion of protein will be _____ (transparent/ translucent).

b.   A colloidal dispersion of protein _____ (will/will not) settle.

c.   A colloidal dispersion of protein _____ (will/will not) pass through a membrane.

174

Many proteins are colloids. Should such proteins be able to pass through the membranes of the kidneys and be present in the urine? _____

Should salt, NaCl? _____

a.  translucent
b.  will not
c.  will not

no; yes

# Diffusion and Osmosis 6

## DIFFUSION

If a bottle of perfume is opened in a room, the odor is soon apparent throughout the room. The molecules of perfume diffuse, or spread out, into the available volume of air.

Next, suppose that a sugar solution is placed on one side of a permeable (nonselective) membrane and that pure water is placed on the other side of that membrane.

○ = sugar molecules

The sugar molecules (the solute) will diffuse across the permeable membrane until the concentrations on both sides of the membrane are equal.

When this equilibrium occurs, sugar (solute) molecules will continue to diffuse across the membrane in both directions so that there will be no net change in concentration.

Diffusion is the tendency to spread into available space. That is, a substance will diffuse from where it is more concentrated to where it is less concentrated. Diffusion across a membrane is called *passive transport* because it requires no external energy.

During diffusion, solute particles move from:

__ high concentration of solute to lower concentration of solute
__ low concentration of solute to higher concentration of solute

**high concentration of solute to lower concentration of solute**

In passive transport, solute particles flow from a _____ (high/low) concentration to one of _____ (higher/lower) concentration.

**high; lower**

Can diffusion occur across a permeable membrane? _____

**yes**

Can diffusion occur without a membrane? _____

**yes**

Do all solutions have the same solute concentration? _____

no

# OSMOSIS

A solution that has a higher solute concentration than another solution is said to be *hypertonic* (hyperosmotic) when compared with the second solution.

A solution that has a lower solute concentration than another solution is said to be *hypotonic* (hypoosmotic) when compared with the second solution.

A solution that has the same solute concentration as another solution is said to be *isotonic* (isoosmotic) when compared with the second solution.

Look at the following illustrations of solutions whose solute is indicated by small circles.

180

a.   Solution A is hypertonic to solution(s) _____.

b.   Solution A is hypotonic to solution(s) _____.

c.   Solution A is isotonic to solution(s) _____.

a.  C, E
b.  B
c.  D

**nonselective**

A permeable membrane is _____ (selective/nonselective).

A semipermeable membrane is one that is selectively permeable.

Consider the following system in which a selectively permeable membrane separates a dilute (weak) sugar solution from a concentrated (strong) sugar solution. We will assume that this semipermeable membrane is permeable to water but not to sugar molecules.

Water will flow through the semipermeable (selectively permeable) membrane and equalize the concentrations. Such a flow is called *osmosis*. The results will be:

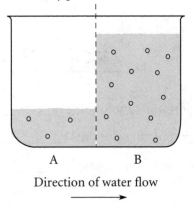

## 182

Osmosis is a type of diffusion. Does it require energy? _____
Is osmosis passive transport? _____

**no; yes**

## 183

Osmosis is the flow of _____ across a membrane.

**water**

## 184

Original solution A is _____ (hypertonic/hypotonic) when compared with original solution B.

**hypotonic**

## 185

Original solution B is _____ when compared with original solution A.

**hypertonic**

## 186

During osmosis, water flows from a hypotonic solution to a hypertonic solution, or from a _____ (weaker/stronger) solution to a _____ (weaker/stronger) solution.

**weaker; stronger**

## 187

Osmosis is the flow of _____ through a selectively permeable membrane.

**water**

yes

### 188

Will osmosis continue until the two solutions have equal concentrations? _____

solute; water molecules

### 189

Diffusion of sugar is the flow of _____ particles; osmosis is the flow of _____ through a selectively permeable (semipermeable) membrane.

hypertonic (concentrated); hypotonic (dilute)

### 190

Diffusion is the flow of solute particles from a _____ solution to a _____ solution.

hypotonic (dilute); hypertonic (concentrated)

### 191

Osmosis is the flow of water through a membrane from a _____ solution to a _____ solution.

yes; yes

### 192

Is diffusion passive transport? _____ Is osmosis? _____

Consider the following system where dilute and concentrated salt solutions are separated by a selectively permeable membrane that allows the salt ions and water to flow through.

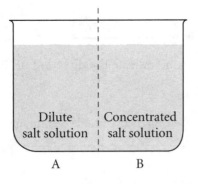

a. What term is used to compare the strength of solution A with that of solution B? _____

b. What term is used to compare the strength of solution B with that of solution A? _____

c. When diffusion takes place, the flow of solute will be from _____ to _____ solution.

a. hypotonic
b. hypertonic
c. hypertonic (concentrated); hypotonic (dilute)

Such a flow of solute is also called _____ transport.

passive

195

While diffusion is taking place, osmosis can also occur.

During osmosis, _____ flows through a membrane from a _____ solution to a _____ solution.

water; hypotonic; hypertonic

no (opposite directions)

Selectively permeable membrane

Does osmosis take place in the same direction through the membrane as diffusion of a solute? _____

In the following diagram, indicate the direction of flow of osmosis and diffusion of ions.

Selectively permeable membrane

## OSMOTIC PRESSURE

198

Consider the following system in which water is placed on both sides of a U-tube that has a selectively permeable membrane (permeable to water but not to solutes) dividing the two halves.

Will osmosis take place? _____ Why or why not? _____

_____

Next, consider a similar U-tube in which water is placed on one side and sugar solution on the other. The membrane is impermeable to sugar but permeable to water.

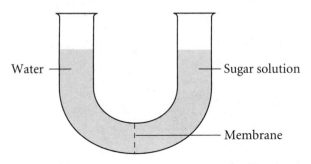

a. Will osmosis take place? _____

b. In which direction will osmosis take place? _____

no; there is no difference in concentration

a. yes
b. into sugar solution

The net result will be a rise in the liquid level on the sugar side and a drop in the liquid level on the water side, as in the following figure.

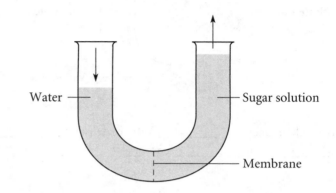

If a piston is placed on the sugar side of the U-tube and pressure is applied to that piston, then osmosis can be prevented. The amount of pressure required to counteract osmosis is called the *osmotic pressure.*

In general, the greater the difference in concentration on two sides of a membrane, the greater the osmotic pressure.

200

C; B

In the following diagrams of sugar solutions separated from each other by a membrane impermeable to sugar but permeable to water, which will exert the greatest osmotic pressure? _____ The least? _____

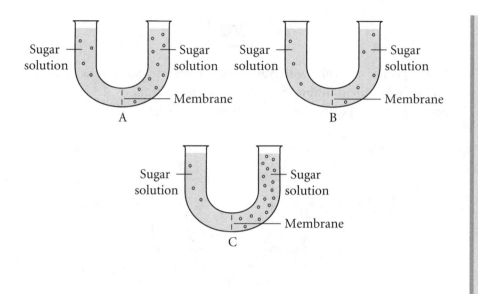

A    B

C

**201**

Consider the following system where a U-tube has water on one side of a membrane and a dilute NaCl solution on the other. The membrane is permeable to sodium and chloride ions and to water.

Pure water

Dilute NaCl solution

Membrane

a.   Will osmosis take place? _____

b.   In which direction? _____

c.   Will diffusion take place? _____

d.   In which direction? _____

e.   Will the two concentrations equalize? _____

a.  yes
b.  into NaCl solution
c.  yes
d.  into the water
e.  yes

a.  yes
b.  yes
c.  yes
d.  yes
e.  **diffusion, or passive
    transport**

Now consider a U-tube with NaCl solution on one side and KCl solution on the other; the solutions have the same concentrations. The membrane is equally permeable to both NaCl and KCl.

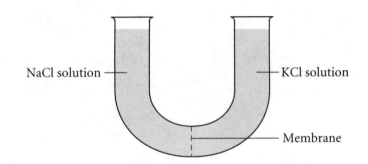

a.  Will NaCl diffuse? _____

b.  Will KCl diffuse? _____

c.  Will the concentrations of KCl be the same on each side when diffusion is finished? _____

d.  Will the concentrations of NaCl be the same on each side?

    _____

e.  What is this motion of solute particles called? _____

## ACTIVE TRANSPORT

yes

203

Next, consider an animal cell with cation concentrations in millimoles per liter as listed in the following table.

| Cation | Intracellular | Extracellular |
|--------|---------------|---------------|
| Na     | 10            | 140           |
| K      | 140           | 4             |

The same situation is shown in the following diagram, where the size of the symbol indicates relative concentration.

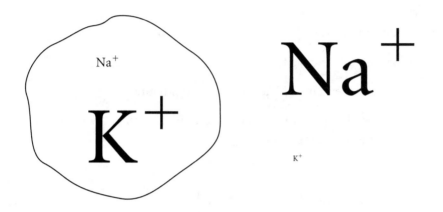

If diffusion or passive transport were the only factor affecting these ions, would the concentrations tend to equalize? _____

204

Because the extracellular and intracellular concentrations of sodium and potassium ions are unequal in animal cells, there must be another force acting (shown in the following diagram).

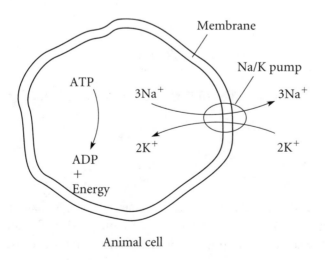

Animal cell

a. ATP
b. $Na^+$
c. $K^+$
d. no
e. 2

This type of movement is called *active transport.*

a.  What energy compound is required for active transport? _____

b.  What ion does active transport pump out of the cell? _____

c.  What ion does active transport pump into the cell? _____

d.  Are the movements of Na$^+$ and K$^+$ equal in amount? _____

e.  For every three sodium ions pumped out of a cell, how many potassium ions are pumped in? _____

**active; ATP**

**205**

This type of ion movement is called _____ transport and requires the energy compound _____.

**no**

**206**

This ion system pumping sodium ions out of a cell and potassium ions into a cell is called a *sodium-potassium pump.*

Are Na$^+$ and K$^+$ pumped equally across a membrane? _____

**3; 2**

**207**

How many Na$^+$ are pumped out? How many K$^+$ are pumped in? _____

**less**

**208**

Because both Na$^+$ and K$^+$ are positively charged ions, 3 positive charges are pumped out of a cell for every 2 that are pumped in. Should the inside of the cell be more positive or less positive than the outside of the cell? _____

## 209

Should the inside of the cell be positive or negative when compared with the outside of the cell? _____

## 210

Animal cells have a *membrane potential* or a small voltage difference between the inside and outside of the cell.

An animal cell has a small voltage difference or _____ between the inside and outside of the cell.

## 211

Complete the following diagram.

_____ transport

Membrane

# Nerve Cells  7

## DEPOLARIZATION
## AND REPOLARIZATION

### 212

A resting neuron has a higher _____ ($K^+$/$Na^+$) concentration inside the cell and a higher _____ ($K^+$/$Na^+$) ion concentration outside that cell.

### 213

A resting cell has a ($+$/$-$) charge inside.

### 214

A resting neuron has a fairly constant electrical potential difference (voltage) of $-50$ to $-100$ millivolts inside. The following diagram represents a resting neuron. Note that the membrane potential is between $-80$ and $-100$ mV. Does a resting neuron have a positive or negative electrical potential inside? _____

$K^+$; $Na^+$

$-$

negative

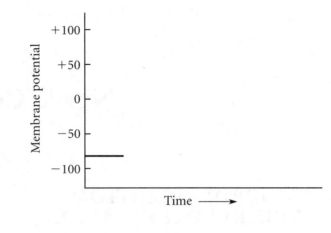

a. closed potassium
   channel
b. open potassium
   channel

215

Some cells, particularly nerve and muscle cells, can generate changes in the electrical potential difference across their cell membranes when a stimulus is applied. Such cells have sodium and potassium channels that can open to allow these ions in or out of the cell. When closed, these channels do not allow ions to pass through.

A sodium channel may be represented by a square shape (S for sodium; S for square). The diagram on the left represents an open sodium channel, and the diagram on the right represents a closed sodium channel.

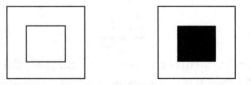

If a triangular shape represents a potassium channel, what do the following diagrams represent? a. _____ b. _____

a                              b

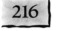

The following diagram illustrates a resting neuron. Which channels are open? _____ Closed? _____

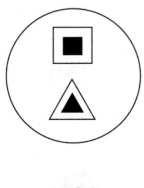

**both Na and K channels are closed**

If both Na and K channels are closed, should the intracellular $Na^+$ and $K^+$ concentrations remain constant?

**yes**

If a neuron is stimulated, then the sodium channels open but the potassium channels remain closed, as in the following diagram.

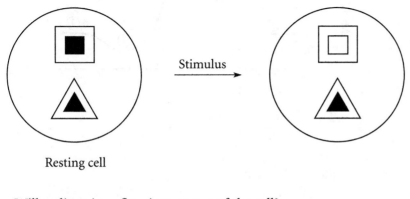

Resting cell

Stimulus

a. Will sodium ions flow into or out of the cell? _____

b. Will potassium ions flow into or out of the cell? _____

**a. into**
**b. neither (the potassium channel is closed)**

+

a stimulus

resting state

Sodium ions have what type of charge, + or −? _____

As sodium ions flow into the cell, the potential of that cell will change from − to +. The cell is said to be *depolarized.*

### 220

What causes the sodium channels to open? _____

### 221

As the cell becomes depolarized, the membrane potential changes from − to +. Section 2 of the following diagram illustrates depolarization. What does section 1 indicate? _____

Color in the following diagram to show which channels should be open and which should be closed. What is effect *A*?

    *A* ⟶

Resting cell                    Depolarization

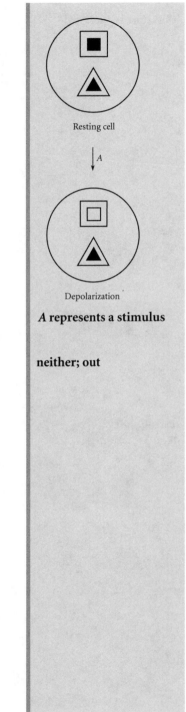

Resting cell

↓ *A*

Depolarization

**A represents a stimulus**

**neither; out**

223

Once the cell has become depolarized, the sodium channels close and the potassium channels open.

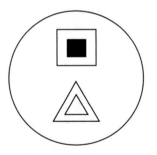

Will sodium ions flow into or out of the cell? _____

Will potassium ions flow into or out of the cell? _____

**it becomes negative**

## 224

This phase is called *repolarization*, as shown in section 3 of the following diagram.

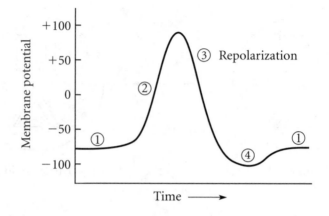

During repolarization, as potassium ions flow out of the cell, what happens to the membrane potential? _____

**resting state;
depolarization**

## 225

What is section 1 called in the problem 224 diagram? _____

What is section 2 called? _____

## 226

Color the following diagram to indicate which of the following channels are open or closed.

Resting cell

Stimulus

Depolarization

Repolarization

Resting cell

Stimulus

Depolarization

Repolarization

The flow of potassium ions out of the cell during repolarization continues afterward because these channels are slow to close. Such continued flow produces an *undershoot* or greater negative charge inside the cell than was present in the resting state, as shown in the following diagram.

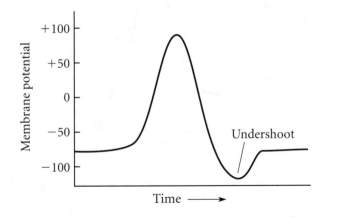

Finally, both channels are closed and active transport takes over to return the cell to its resting state.

## 227

a.  sodium
b.  potassium
c.  3 Na$^+$ out; 2 K$^+$ in
d.  potassium
e.  sodium

a.  During active transport, _____ ions are carried out of the cell.

b.  During active transport, _____ ions are carried into the cell.

c.  What are the relative number of Na$^+$ and K$^+$ ions moved out of and into the cell? _____

d.  During undershoot, which channels are slow to close? _____

e.  During undershoot, which channels remain closed? _____

## 228

1: resting state
2: depolarization
3: repolarization
4: undershoot
5: resting state

In the following diagram, label sections 1 through 5.

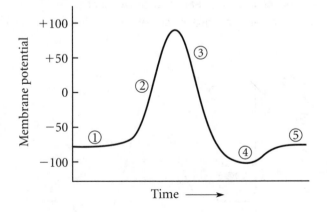

229

In the following diagram, indicate the charges inside the cell membrane. Also mark which channels are open and which are closed. Finally, label *A* and *B*.

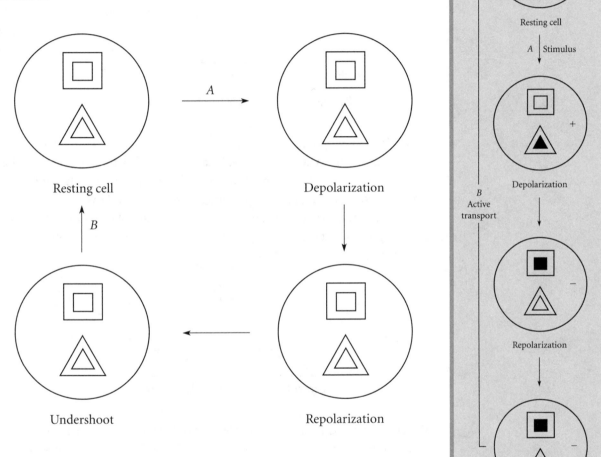

Resting cell

A

Depolarization

B

Undershoot

Repolarization

Resting cell

*A* | Stimulus

Depolarization

*B*
Active
transport

Repolarization

Undershoot

# PROPAGATION OF A NERVE IMPULSE

Nerve impulses are propagated in an axon by means of a series of depolarizations and repolarizations, such as indicated in the following drawing.

```
+ + + +        – – – –        + + + +
–––––          + + + +        – – – –

– – – –        + + + +        – – – –
+ + + +        – – – –        + + + +
```

Resting state    Depolarization    Repolarization

How does this propagation occur? Depolarization increases the membrane potential to a critical level called the threshold potential. This increase in turn triggers an *action potential*, or nerve impulse, which is transmitted to an adjacent segment. Consider the following diagram of the cross section of a small section of an axon that illustrates three adjacent segments in their resting state with a negative potential inside each segment.

```
+   +   +   +   +   +   +   +   +
–   –   –   –   –   –   –   –   –

–   –   –   –   –   –   –   –   –
+   +   +   +   +   +   +   +   +
‿‿‿‿‿      ‿‿‿‿‿      ‿‿‿‿‿
Segment A   Segment B   Segment C
```

When a stimulus is applied to segment A, it causes that segment to become depolarized or positively charged in its inner membrane, which produces an action potential.

```
–   –   –   +   +   +   +   +   +
+   +   +   –   –   –   –   –   –

+   +   +   –   –   –   –   –   –
–   –   –   +   +   +   +   +   +
‿‿‿‿‿      ‿‿‿‿‿      ‿‿‿‿‿
Segment A   Segment B   Segment C
Depolarized   Resting state   Resting state
```

The depolarization of segment A in turn causes the depolarization of segment B so that the action potential (nerve impulse) travels along the indicated path.

| + + + | − − − | + + + |
| --- | --- | --- |
| − − − | + + + | − − − |
| − − − | + + + | − − − |
| + + + | − − − | + + + |

| Segment A | Segment B | Segment C |
| --- | --- | --- |
| Repolarized | Depolarized | Resting state |

—— Impulse ⟶

## 230

While the impulse is traveling from A to B, segment A becomes repolarized and returns to its resting state.

The depolarization of segment B will have what effect on segment C?

_____

it will become depolarized

## 231

What happens to the action potential (nerve impulse)? _____

_____

it travels from segment B to segment C

## 232

What happens to segment B after the impulse is transmitted to C? _____

_____

it becomes repolarized and then returns to its resting state

This effect may be considered to be similar to that of a section of standing dominoes. When the first one is pushed over, the motion is transmitted to succeeding ones just as a nerve impulse travels from one end of an axon to the other.

Note, however, that the last falling domino is too far from the next one to cause it to fall. The effect then stops.

Similarly, the movement of a nerve impulse travels along an axon to the end but cannot jump across the gap to the next axon. Something must carry that impulse across the gap.

The following diagram illustrates the end of one axon and the beginning of the succeeding one.

Can the nerve impulse jump across the synaptic cleft? _____

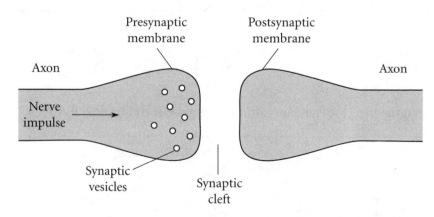

no

## 234

When the action potential reaches the presynaptic membrane, it depolarizes it. The depolarization of the presynaptic membrane triggers an influx of calcium ions, which in turn causes the synaptic vesicles to fuse with the presynaptic membrane.

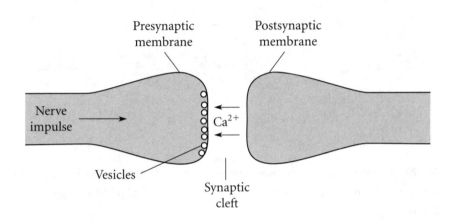

What effect does a nerve impulse have on the presynaptic membrane?

_____

## 235

The depolarization of the presynaptic membrane causes the influx of which ions? _____

## 236

The influx of calcium ions has what effect on the vesicles?

_____

When the vesicles fuse to the presynaptic membrane, they release neurotransmitters that can travel across the synaptic cleft to the postsynaptic membrane.

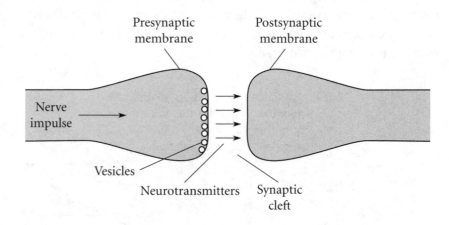

At the postsynaptic membrane, the neurotransmitters bind to receptor sites, which then open specific ion channels in the postsynaptic membrane.

yes

When an ion channel is open, can ions move across a membrane?
_____

no

When an ion channel is closed, can ions move across a membrane?
_____

neurotransmitter

239

When the vesicles bind to the presynaptic membrane, what type of substance is released? _____

## 240

When a neurotransmitter is released, where does it go? _____

into the synaptic cleft

## 241

When the neurotransmitter reaches the postsynaptic membrane, what effect does it have on receptor sites? _____

it causes ion channels to open

Once the ion channels open (which channels depend on the receptor molecules associated with those channels), selected ions enter the postsynaptic membrane, causing depolarization and thus a new action potential there. The new action potential then carries the nerve impulse along that axon.

## 242

Does the identity of the neurotransmitter determine which ion enters the postsynaptic membrane? _____

yes

## 243

When the ion channels are open, selected ions can enter the postsynaptic membrane, causing _____.

depolarization

## 244

When the postsynaptic membrane is depolarized, what effect may be produced? _____

an action potential

it begins sending a nerve impulse along that axon

## 245

What effect does this action potential have on the second axon?

After the postsynaptic membrane is depolarized, enzymes degrade the neurotransmitters so that the membrane is ready for succeeding signals from new molecules of neurotransmitters.

A: presynaptic membrane
B: synaptic cleft
C: postsynaptic membrane
D: vesicles
E: ion channels
F: action potential
G: flow of neurotransmitters

## 246

Label *A* through *G* in the following diagram.

PART **II**

Organic
Chemistry

# The Covalent Bond 8

247

You've already learned that atoms tend to lose electrons to form positively charged ions if they have 1, _____, or _____ electrons in their valence shells.

On the other hand, atoms that have _____ or _____ electrons in their valence shells tend to gain electrons to form _____ charged ions.

2; 3; 6; 7; negatively

248

Atoms with 4 or 5 electrons in their valence shells tend to *share* electrons. When atoms share electrons, they form a *covalent bond*.

Which of the following illustrated atoms would form covalent bonds? _____

C and N

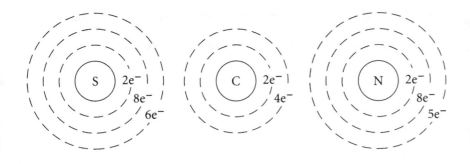

a. Mg, H, Al
b. Cl, S
c. Ne
d. **stable and unreactive**

249

Answer the following questions about these illustrated atoms.

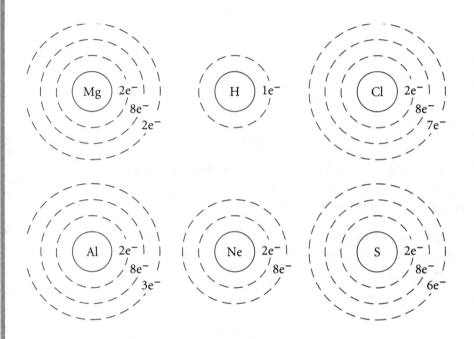

a.  Which of the atoms will lose electron(s) to form positively charged
    ions? _____

b.  Which of the atoms will gain electron(s) to form negatively charged
    ions? _____

c.  Which of the atoms has a complete valence shell? _____

d.  You know that elements with a complete valence shell are usually:

    __ unstable and reactive      __ stable and unreactive

a. covalent
b. nonelectrolyte

a. When two atoms share electrons, they are held together by a
_____ bond.

b. When two atoms share electrons, no ions are produced. Is the
resulting compound an electrolyte or a nonelectrolyte? _____

ionic

Atoms with 4 or 5 electrons in their valence shells tend to share electrons,
but under some conditions atoms with more or fewer electrons in their
valence shells may also share electrons.

When one atom loses an electron and another atom gains that
electron, the ions thus formed are held together by a(n) _____
(covalent/ionic) bond.

a. electrolytes
b. nonelectrolytes

a. Compounds containing ionic bonds are _____
(electrolytes/nonelectrolytes).

b. Compounds containing only covalent bonds are _____
(electrolytes/nonelectrolytes).

4

## 253

The carbon atom, atomic number 6, has 4 electrons in its valence shell. It thus needs _____ more electrons to become stable. It can get these electrons by sharing. The Lewis dot structure for carbon is

2

## 254

When atoms share electrons, they always try to reach stable configurations of 8 electrons in the valence shells. The only exception to this rule (the octet rule) is the hydrogen atom. It reaches a stable configuration when it has 2 electrons in its valence shell, which is also its only energy level. This first energy level can hold only _____ electrons.

4

## 255

Consider the carbon atom with 4 electrons in its valence shell. With how many hydrogen atoms can it share electrons to reach a stable configuration of 8 electrons in its valence shell? _____

## 256

The four hydrogen atoms can share electrons with a carbon atom to form a compound of the following type:

$$
\begin{array}{c}
\text{H} \\
\vdots \\
\text{H} : \text{C} : \text{H} \\
\vdots \\
\text{H}
\end{array}
\quad \text{or} \quad
$$

There are _____ electrons around the carbon atom and _____ electrons around each hydrogen atom.

## 257

When two hydrogen atoms combine, they share electrons. Each hydrogen atom then has _____ electrons around it.

H $\cdot\cdot$ H

## 258

The diagram illustrates two hydrogens held together by a covalent bond. Indicate by means of an arrow the location of the covalent bond in the compound.

H : H

**4**

**7**

. Cl :
⠒⠒

**covalent**

## 259

In the following compound, what is the total number of covalent bonds? _____

<pre>
        H
        ··
  H :  C  : H
        ··
        H
</pre>

## 260

The chlorine atom, atomic number 17, has _____ electrons in its valence shell.

## 261

Show the arrangement of dots for the Lewis dot structure for chlorine.

Cl

When two chlorine atoms combine, they share electrons to reach a stable configuration of **8** electrons in the valence shell of each atom, or

## 262

What type of bond is there between two chlorine atoms? _____

8;8

The Lewis dot structure for $CO_2$ is

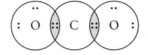

How many electrons are there around the carbon atom? _____
Around each oxygen atom? _____

2 pair

How many shared electron pairs are there between the carbon atom and
*each* oxygen atom? _____
Such bonds are called *double covalent bonds.*

4; 2

A double covalent bond contains a total of how many electrons? _____
A (single) covalent bond contains a total of how many electrons? _____

2

The $CO_2$ molecule contains how many double bonds? _____

covalent (single cova-
lent); triple covalent

**267**

The compound

$$H : C :: N :$$

has what type of bond between the carbon and the hydrogen?
_____

Between the carbon and the nitrogen? _____

structure (b) because it
shows 2 electrons around
each hydrogen, 8 around
the carbon, and 8 around
the nitrogen

Cl — Cl

**268**

Using the octet rule, predict which of the following structures would
best represent the compound $CH_3NH_2$.

a.     H           b.     H          c.     H

**269**

A covalent bond is frequently indicated by a short line rather than
by dots; thus, the hydrogen molecule may be represented as H : H
or H — H. The short line indicates a pair of shared electrons.

The compound $CH_4$ may be represented as

Give the second representation of the chlorine molecule, $Cl_2$.

$$: Cl : Cl : \quad or$$

### 270

Give the line representation of the bonds in HCN whose Lewis dot structure is given in frame 267.

### 271

a.  The carbon atom, atomic number 6, needs how many electrons to complete its valence shell? _____

b.  Because the carbon atom has 4 electrons in its valence shell, and because it will tend to share these electrons, how many covalent bonds will a carbon atom form? _____

**a.  4**
**b.  4**

The carbon atom, therefore, must have four covalent bonds, or must have four bonds attached to it. These bonds may be indicated as follows:

Note that each carbon atom has four bonds attached to it, regardless of how the bonds are arranged.

### 272

The hydrogen atom has 1 electron in its outer energy level, the first energy level. When the hydrogen atom shares electrons, how many more does it need to complete that valence shell? _____

1

a. 6;
   :Ö:
   ̈

b. 2

c. 2

H
|
H—C—H
|
H

a. 3
b. 3

## 273

a. The oxygen atom, atomic number 8, has _____ electrons in its valence shell. Its Lewis dot structure is _____.

b. When the oxygen atom shares electrons, how many more electrons does it need to complete its valence shell? _____

c. Therefore, each oxygen atom must have how many bonds attached to it? _____

## 274

Draw the structure of the compound formed when four hydrogen atoms form bonds with a central carbon atom.

C

## 275

The arrangement of carbon atoms C — C indicates a covalent bond or a shared electron pair between two carbon atoms.

a. How many hydrogen atoms may be attached to the left carbon atom? _____

b. How many hydrogen atoms may be attached to the right carbon atom? _____

## 276

a.　Diagram all the hydrogen atoms attached to this structure:

C — C

b.　How many bonds does each carbon atom have? _____

c.　How many bonds does each hydrogen atom have? _____

## 277

Consider this arrangement of carbon atoms:

C — C — C

a.　How many hydrogen atoms may be attached to the left carbon atom? _____

b.　How many hydrogen atoms may be attached to the center carbon atom? _____

c.　How many hydrogen atoms may be attached to the right carbon atom? _____

## 278

Diagram the compound containing three attached carbon atoms, indicating all the hydrogen atoms connected to those carbon atoms.

C — C — C

a.

b. 4
c. 1

a. 3
b. 2
c. 3

a. **3**
b. **3**
c. **3**
d. **3**
e. **0 (or none)**

a.

b. **it already has four bonds**

Consider this arrangement of carbon atoms:

$$
\begin{array}{c}
\text{C} \\
| \\
\text{C} - \text{C} - \text{C} \\
| \\
\text{C}
\end{array}
$$

a.  How many hydrogen atoms may be attached to the right carbon atom? _____

b.  How many hydrogen atoms may be attached to the left carbon atom? _____

c.  How many hydrogen atoms may be attached to the upper carbon atom? _____

d.  How many hydrogen atoms may be attached to the lower carbon atom? _____

e.  How many hydrogen atoms may be attached to the center carbon atom? _____

a.  Diagram the structure of the compound containing the following arrangement of carbon atoms, indicating all the hydrogen atoms.

$$
\begin{array}{c}
\text{C} \\
| \\
\text{C} - \text{C} - \text{C} \\
| \\
\text{C}
\end{array}
$$

b.  Why can there be no hydrogen atoms attached to the center carbon atom?_____

## 281

Consider this arrangement of carbon atoms:

$$C = C$$

There are two bonds (a double bond) between the carbon atoms. Each bond (a single bond) represents 1 pair of shared electrons. A double bond represents _____.

2 pairs of shared electrons

## 282

a.  To become stable, a carbon atom must form how many covalent bonds? _____

b.  In the structure $C = C$, how many hydrogen atoms may be attached to the left carbon atom? _____

c.  In the space provided, diagram the structure of the compound containing two carbon atoms connected by a double bond. Show all the hydrogen atoms.

a.  4

b.  2

c.

When you have completed your diagram, check it by counting the number of bonds around each carbon atom.

a. 2
b. 1
c. 3

## 283

Consider this arrangement of carbon atoms:

a.  How many hydrogen atoms may be attached to the left carbon atom? _____

b.  How many hydrogen atoms may be attached to the center carbon atom? _____

c.  How many hydrogen atoms may be attached to the right carbon atom? _____

## 284

Complete the structure of the following compound, showing all the hydrogen atoms.

## 285

Diagram the structure of carbon tetrachloride, $CCl_4$, in which each chlorine atom has one bond attached to it.

### 286

Diagram the structure of chloroform, $CHCl_3$.

### 287

Nitrogen, atomic number 7, has 5 electrons in its valence shell and so must have three bonds attached to it. Diagram the ammonia molecule, $NH_3$.

Note that the structure

may be simplified and written as $CH_3CH_2CH_2CH_3$. Likewise,

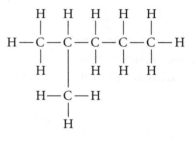

may be simplified and written as $CH_3CH(CH_3)CH_2CH_2CH_3$, and

may be simplified and written as $CH_3CH=CHCH_3$.

## 288

How may the following structures be written in simplified form?

a.

$$H-\underset{\underset{H}{|}}{\overset{\overset{H}{|}}{C}}-\underset{\underset{H}{|}}{\overset{\overset{H}{|}}{C}}-H$$

_____

b.

$$H-\underset{\underset{H}{|}}{\overset{\overset{H}{|}}{C}}-\underset{\underset{H}{|}}{\overset{\overset{H}{|}}{C}}-\underset{\underset{\underset{H}{|}}{\overset{|}{C}}}{\overset{\overset{H}{|}}{C}}-\underset{\underset{H}{|}}{\overset{\overset{H}{|}}{C}}-\underset{\underset{H}{|}}{\overset{\overset{H}{|}}{C}}-H$$

$$H-\underset{\underset{H}{|}}{C}-H$$

_____

c.

$$H-\underset{\underset{H}{|}}{\overset{\overset{H}{|}}{C}}-C\equiv C-H$$

_____

d.

$$H-\underset{|}{\overset{\overset{H}{|}}{C}}-H$$

$$H-\underset{\underset{H}{|}}{\overset{\overset{H}{|}}{C}}-\underset{|}{C}-\underset{\underset{H}{|}}{\overset{\overset{H}{|}}{C}}-\underset{\underset{H}{|}}{\overset{\overset{H}{|}}{C}}-H$$

$$H-\underset{\underset{H}{|}}{\overset{|}{C}}-H$$

_____

a.  $CH_3CH_3$
b.  $CH_3CH_2CH(CH_3)CH_2CH_3$
c.  $CH_3C\equiv CH$
d.  $CH_3C(CH_3)_2CH_2CH_3$

The structure for benzene, $C_6H_6$, may be represented in two different ways, depending on the arrangement of the electrons:

Such structures are called *resonance* structures and are indicated by a double-headed arrow.

Often, either of these resonance structures is used to represent benzene. For simplicity, however, benzene is usually abbreviated as

289

What is the abbreviated structure for the following?

a.

a.

b.

Draw the complete structure for each of the following simplified formulas:

a. $CH_3CH_2CH_3$

b. $CH_3CH=CHCH_3$

c.

Cl

d. $CH_3CH(CH_3)CH_2CH_2CH_3$

# Polar and Nonpolar Covalent Bonds  9

## 291

Electronegativity is the attraction of an atom for electrons. The following table indicates the electronegativities of several elements.

| F  | 4.0 | Br | 2.8 | Al | 1.5 |
|----|-----|----|-----|----|-----|
| O  | 3.5 | C  | 2.5 | Ca | 1.0 |
| Cl | 3.1 | H  | 2.1 | Na | 0.9 |
| N  | 3.0 |    |     |    |     |

Note that nonmetals have higher electronegativities than do metals. The difference in atoms' electronegativities determines the type of bond that can form between them.

When atoms of two nonmetals combine, the difference in electronegativities is small or zero and a covalent bond results.

When atoms of a metal and a nonmetal react, the difference in electronegativities is relatively larger and an ionic bond results.

a. Which of these elements has the greatest attraction for electrons?

_____

b. Which of these elements has the least attraction for electrons?

_____

a. F
b. Na

## 292

In the compound Cl — Cl, or Cl:Cl, there is a pair of shared electrons between the chlorine atoms. Is there a difference in electronegativity between the two chlorine atoms? _____

no

a. Cl
b. closer to

## 293

Because the two chlorine atoms have the same electronegativity, the electrons should be shared equally between them. That is, the electrons should not be closer to one chlorine atom than to the other. Such a bond is called a *nonpolar covalent bond*, and such a compound is called a *nonpolar compound*. Nonpolar means that there is no negative end and no positive end to the bond or to the molecule.

a.  In the covalent compound HCl, H — Cl, or H:Cl, which element has the greater electronegativity? _____

b.  Because Cl is more electronegative than H, the shared electrons should be _____ (closer to/farther from) the chlorine.

a. yes
b. polar
c. polar

## 294

Because the shared electrons are closer to the Cl than to the H, the Cl will have a partial negative charge and the H will have a partial positive charge. This property may be indicated by a $\delta$ (Greek delta) sign.

$$\overset{\delta+ \;\; \delta-}{HCl}$$

Note that it is a partial charge, not an ionic charge.

a.  Is one end of the molecule more negative than the other end?

_____

b.  The covalent bond between the H and the Cl is _____ (polar/nonpolar).

c.  Is the HCl molecule polar or nonpolar? _____

nonpolar

## 295

Consider the covalent compound XY. If both atoms have the same electronegativity, will the covalent bond between them be polar or nonpolar? _____

If there is a difference in electronegativity between X and Y, then the covalent bond between them will be _____ and the compound XY will be _____.

polar; polar

Now consider the covalent compound $CO_2$, or O::C::O

a. Which atom is more electronegative? _____ Which is less electronegative? _____

b. Which atom will have a partial + charge? _____ Which will have a partial − charge? _____

a. O; C
b. C; O

The $CO_2$ molecule may be represented as a linear molecule.

$$\overset{\delta-}{O}::\overset{\delta+}{C}::\overset{\delta-}{O}$$

Both ends of the $CO_2$ molecule have a partial _____ (positive/negative) charge.

negative

**299**

a. Does the $CO_2$ molecule have a positive and a negative end? _____

b. Is $CO_2$ a polar or nonpolar compound? _____

a. no
b. nonpolar

nonpolar

In general, if a covalent compound is symmetrical, it will be _____ (polar/nonpolar).

H; O

In the $H_2O$ molecule, which contains covalent bonds, which atom has a partial positive charge? _____

Which has a partial negative charge? _____

b; a

The structure of the water molecule may be represented as either

a.   $\overset{\delta+}{H}—\overset{\delta-}{O}—\overset{\delta+}{H}$     or     b.   $\overset{\delta+}{H}—\overset{\delta-}{O}$
$\phantom{a.   H—O—H     or     b.   H—O}\diagdown$
$\phantom{a.   H—O—H     or     b.   H—O\diagdown}H^{\delta+}$

The structure in a is theoretical; it does not exist.

Which structure, a or b, represents a polar compound? _____

Which represents a nonpolar compound? _____

b

The water molecule has been found to be a polar molecule. Which of the structures in problem 302 more accurately depicts the water molecule?

_____

## 304

One general rule of solubility is that *like dissolves like.* That is, polar liquids dissolve polar compounds, and nonpolar liquids dissolve nonpolar compounds. Is water a polar or a nonpolar liquid? _____

## 305

a.  Should NaCl, a polar (ionic) compound, dissolve in water? _____

b.  Should benzene, a nonpolar compound, dissolve in water? _____

c.  Should acetone, a nonpolar liquid, dissolve in benzene, a nonpolar liquid? _____

polar

a. yes
b. no
c. yes

# Functional Groups in Organic Compounds 10

Functional groups impart certain sets of properties to compounds. One example of such a functional group is the —*OH* group.

### 306

Groups that impart certain sets of properties to organic compounds are called _____ groups.

functional

## THE HYDROXYL GROUP

### 307

The —OH group in an ionic compound is called a hydroxide ion.

$$NaOH \longrightarrow Na^+ + OH^-$$
Ionic

The —OH group in a covalent compound is called a *hydroxyl* group; it does not ionize.

Covalent, not ionized

Is a solution of NaOH an electrolyte? _____

yes

no

---

**308**

Consider this compound:

$$H-\underset{\underset{H}{|}}{\overset{\overset{H}{|}}{C}}-OH$$

Is a solution of this compound an electrolyte? _____

The —OH group has the bonds —O—H. It is frequently written as —OH; the bond between the hydrogen and the oxygen is understood and is not written. Note that the oxygen atom still has two bonds attached to it.

In general, covalent compounds containing one or more —OH groups are called *alcohols*.

---

all of them

**309**

Which of the compounds a through e contains a hydroxyl group?

_____

c.   $CH_3CH(OH)CH_3$

d.   $CH_3CH_2CH_2CH(CH_3)CH_2OH$

e.

yes

Glycerol, $C_3H_8O_3$, is one of the primary constituents of fats. Its structure is

Does glycerol contain hydroxyl groups? _____

---

311

hydroxyl; alcohol

The names of alcohols are written with the ending *ol*. $CH_3OH$ is methanol. $CH_3CH_2OH$ is ethanol.

If the name of a compound ends in *ol*, that compound contains a _____ group and is called a(n) _____.

---

312

a, b, e

Which of the following compounds is(are) an alcohol? _____

a.  cholesterol

b.  tocopherol (vitamin E)

c.  thiamine (vitamin $B_1$)

d.  cortisone

e.  retinol (vitamin A)

---

313

hydroxyl

Alcohols contain which functional group? _____

# THE CARBONYL GROUP

A second functional group is the *carbonyl* group $C = O$.

**yes**

### 314

Look at the following structures:

$$\textbf{a.} \; CH_3CH_2CH_2\overset{\displaystyle H}{\underset{\displaystyle |}{C}} = O \quad \text{and} \quad \textbf{b.} \; CH_3\overset{\displaystyle}{\underset{\displaystyle \| \\ O}{C}} \, CH_2CH_3$$

Do both contain a carbonyl group? _____

**aldehyde; ketone**

### 315

In compound a, the carbonyl group is at the end of the carbon chain; in compound b, the carbonyl group is not at the end of the carbon chain.

Compounds in which the carbonyl group is at the end of the carbon chain are called *aldehydes*.

Compounds in which the carbonyl group is not at the end of the carbon chain are called *ketones*.

Compound a is an example of a(n) _____ (aldehyde, ketone).

Compound b is an example of a(n) _____ (aldehyde, ketone).

**carbonyl**

### 316

Note that the functional group of an aldehyde may be written as

$$CH_3CH_2CH_2\overset{\displaystyle H}{\underset{\displaystyle |}{C}} = O \quad \text{or} \quad CH_3CH_2CH_2CHO$$

and the functional group of a ketone may be written as

$$CH_3\overset{\displaystyle}{\underset{\displaystyle \| \\ O}{C}}CH_2CH_3 \quad \text{or} \quad CH_2COCH_2CH_3$$

The $C = O$ functional group is called the _____ group.

## 317

Both aldehydes and ketones contain which functional group?

_____

carbonyl

## 318

In an aldehyde, the carbonyl group is where on the carbon chain?

_____

at the end

## 319

In a ketone, where is the carbonyl group located? _____

not at the end of the chain

## 320

Aldehydes have names ending in *al;* ketones have names ending in *one.*

Compound a in problem 314 is called *butanal.* The ending *al* indicates that butanal is what type of compound? _____

What functional group does it contain? _____

aldehyde; carbonyl

## 321

Compound b in problem 314 is called *butanone.*

Butanone is what type of compound? _____

What functional group does it contain? _____

ketone; carbonyl

The following compounds contain which functional group(s)?

hydroxyl

carbonyl

carbonyl

carbonyl

**322**

$$H-C-C-C-C-OH$$

**323**

$$H-C-C=O$$

**324**

$$H-C-C-C-H$$

**325**

$$\bigcirc\!\!-CHO$$

hydroxyl

_____

Identify the functional groups in the following compounds and identify the type of compound as alcohol, aldehyde, or ketone.

_____

hydroxyl; alcohol

_____

carbonyl; aldehyde

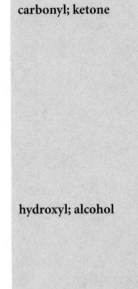

carbonyl; ketone

hydroxyl; alcohol

carbonyl; ketone

carbonyl; ketone

329

_____

330

$$CH_3CH(OH)CH_3$$

_____

331

_____

332

_____

**333**

$$CH_3-C-CH_2-CH_2-CHO$$
$$\overset{\|}{O}$$

carbonyls; aldehyde
and ketone

_____

**334**

hydroxyl; carbonyl
alcohol and aldehyde

_____

**335**

carbonyls; aldehyde
and ketone

_____

**336**

ketones: a, d
alcohols: c, e, g
aldehydes: b, f, h

The names of most ketones end in *one*. An example of a ketone is
acetone. Which of the following substances are ketones? _____
Alcohols? _____ Aldehydes? _____

a.  corticosterone      b.  pyridoxal (a B-vitamin)
c.  cortisol            d.  androsterone
e.  estriol             f.  pentanal
g.  pregnandiol         h.  retinal

# THE CARBOXYL GROUP

acid

### 337

A third functional group is the *carboxyl* group, —COOH or

The carboxyl group ionizes as follows:

Because the carboxyl group yields H$^+$ ions, it is a(n) _____.

Because organic acids contain a carboxyl group, they are also called *carboxylic acids*.

a.  an aldehyde; carbonyl
b.  a carboxylic acid; carboxyl

### 338

The —COOH group may also be written as

a.   A compound with a —CHO group is called a(n) _____ and contains which functional group? _____

b.   A compound with a —COOH group is called a(n) _____ and contains which functional group? _____

Citric acid, α-ketoglutaric acid, succinic acid, and oxaloacetic acid are a part of the Krebs cycle, the human body's primary metabolic cycle. What functional group should they all have in common? _____

carboxyl, or —COOH

### 340

What functional groups are present in the following compounds?

a. pyruvic acid, produced during the oxidation of glucose _____

$$CH_3 - C - COOH$$
$$\overset{\|}{O}$$

b. lactic acid, formed during the fermentation of milk _____

$$CH_3 - CH - COOH$$
$$\underset{OH}{|}$$

c. tartaric acid, found in grapes _____

a. carbonyl and carboxyl (organic acid)
b. hydroxyl and carboxyl
c. hydroxyl and carboxyl

# THE AMINO GROUP

Another functional group is the amino group —$NH_2$

Compounds containing an amino group are called *amines*.

### 341

Consider the following compound:

$$CH_3NH_2$$

Methylamine

Does it contain an amino group? _____

Is it an amine? _____

yes; yes

### 342

Consider the compound glycine:

$$\begin{array}{c} COOH \\ | \\ CH_2NH_2 \end{array}$$

What functional group(s) does it contain? _____

amino and carboxyl

### 343

Is glycine an amine? _____ An acid? _____

yes, yes

Compounds that contain both an amino group and a carboxyl group are called *amino acids.*

**all of them**

## 344

Which of the following compounds contain(s) an amino group?

_____

a. $CH_3NH_2$

b.

$-CH_2CH_2NH_2$

c. $CH_3CHCH_2CH_3$
       |
      $NH_2$

d. $CH_2-CH-CH_2-CH_3$
    |      |
   $NH_2$  $NH_2$

## 345

a. —COOH (carboxyl)<br>
b. an amine<br>
c. an alcohol<br>
d. an aldehyde<br>
e. a ketone

a. An organic acid is a compound containing a(n) _____ group.

b. A compound containing an $—NH_2$ group is called _____.

c. A compound containing an $—OH$ group is called _____.

d. A compound containing a $—CHO$ group is called _____.

e. A compound containing a $C—C—C$ group is called _____.
                                      ‖
                                      O

Identify the functional groups in the following problems.

amino

hydroxyl and carbonyl

hydroxyl and carbonyl

amino and hydroxyl

346

347

348

349

amino and carboxyl

351

carboxyl and hydroxyl

352

amino and carboxyl
(amino acid)

$CH_3CH(NH_2)COOH$

353

carbonyl and amino

$CH_3CCH_2CH_2NH_2$
$\overset{\|}{O}$

354

hydroxyl and carboxyl

$CH_3CH_2CH(OH)COOH$

# THE SULFHYDRYL GROUP

The —SH group is called the sulfhydryl group. A compound containing a sulfhydryl group is called a *thiol*.

**sulfur and hydrogen**

355

The sulfhydryl group contains which elements?

**yes**

356

Consider the following compound:

$$CH_3CH_2SH$$

Ethanethiol

Does it contain a sulfhydryl group? _____

**carboxyl, amino, and
sulfhydryl**

357

Cysteine, a non-essential amino acid has the following structure:

$$
\begin{array}{c}
COOH \\
| \\
NH_2-C-H \\
| \\
CH_2-SH
\end{array}
$$

Cysteine

Which functional groups does it contain? _____

# DISULFIDE BRIDGES

Disulfide bridges contain a —S—S— group.

## 358

What does the term *disulfide* refer to?

two sulfides

The following compound, cystine, contains a disulfide bridge.

Cystine

## 359

2; sulfur; 2; sulfur

The prefix *di* indicates the number _____. Sulfide refers to the element _____. So, a disulfide bridge contains _____ (how many?) atoms of the element _____.

## 360

disulfide; 3

Note the following abbreviated structure of human insulin.

What type of bonds or linkages holds the A and B chains together?

_____

How many bonds of this type are present in human insulin? _____

# THE PHOSPHATE GROUP

The —$OPO_3H_2$ group is called a *phosphate* group. This group may also be written as

361

Compounds containing a phosphate group are called phosphates. Is the following compound a phosphate?

362

In ATP, adenosine triphosphate, the P indicates which functional group?
_____

363

The prefix *tri* indicates how many phosphate groups as being present in ATP? _____

yes

phosphate

3

### 364

Look at the following ladder representation of DNA. The S along the backbone of the chain indicates a sugar. The P indicates which functional group? _____

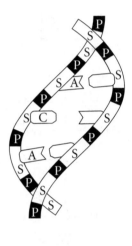

### 365

The following structure represents adenosine monophosphate (AMP). Draw a circle around the phosphate group. This compound also contains an amino and two hydroxyl groups. Draw a square around the amino group and a rectangle around the hydroxyl groups.

# SUMMARY OF
# FUNCTIONAL GROUPS

Name the functional groups present at the following compounds.

**carbonyl and hydroxyls**

Glucose

**amino and carboxyl
(amino acid)**

367

$CH_3CH(NH_2)COOH$

Alanine

CHO

HO — ⎡ ⎤ — CH₂OH
H₃C — ⎣ N ⎦

Pyridoxal
(one of the B-vitamins)

carbonyl and hydroxyl

---

369

Prostaglandin

two hydroxyl groups, a
carboxyl group, and a
carbonyl group

---

370

COOH
|
CHNH₂
|
CH₂OH

Serine

a hydroxyl group, an
amino group, and a
carboxyl group

carbonyl, hydroxyl, and phosphate groups

371

$$CHO$$
$$HC-OH$$
$$H_2C-OPO_3H_2$$

Glyceraldehyde-3-P

carboxyl, amino

372

$$COOH$$

$$NH_2$$

a sulfhydryl group

373

$$CH_3CH_2CH_2SH$$

two carboxyl groups, two amino groups, and one disulfide bridge

374

$$COOH \qquad COOH$$
$$NH_2CH \qquad NH_2CH$$
$$H_2C-S-S-CH_2$$

| 375 |

a carboxyl group, an amino group, and a sulfhydryl group

| 376 |

The following diagram illustrates part of the shape of a protein molecule.

What type of bond or bridge determines the conformation of such a protein? _____

disulfide bridge

What type of functional group(s) is(are) present in each of the following compounds?

| 377 |

*alpha*-ketoglutaric acid _____

carbonyl and carboxyl

| 378 |

methanethiol _____

sulfhydryl

**amino and carboxyl**

**carbonyl**

**carboxyl**

**amino and carboxyl**

**hydroxyl**

**carbonyl**

379

essential amino acids _____

380

acetone _____

381

*trans*-fatty acids _____

382

*para*-aminobenzoic acid _____

383

pregnanediol _____

384

testosterone _____

385

Consider two water molecules whose structures and partial charges are

$$\overset{\delta+}{H}\!\!-\!\!\overset{\delta-}{O} \quad \text{and} \quad \overset{\delta+}{H}\!\!-\!\!\overset{\delta-}{O}$$
$$\underset{H^{\delta+}}{\diagdown} \qquad\qquad \underset{H^{\delta+}}{\diagdown}$$

Which partial charges should attract each other? _____

the negative charge of one water molecule and the positive charge of the other

weak

## 386

Water molecules can form a weak bond between the H of one molecule and the O of another. Such a bond, called a *hydrogen bond*, is caused by the attraction of the partially positive H of one water molecule and the partially negatively charged O (more electronegative) of another water molecule. Hydrogen bonds may be indicated by dotted lines:

Is a hydrogen bond a strong or a weak bond? _____

hydrogen bond

## 387

If many water molecules are bonded together in a similar manner, what type of bond should hold them together? _____

## 388

Hydrogen bonds can also occur between the H attached to an amino group and an O that is part of a carboxyl group. Such hydrogen bonds are present in proteins and in DNA.

Note the following diagram, which illustrates a very small part of a protein molecule. (R represents a carbon chain.)

Pleated sheet structure of a protein

What type of bond holds the sections together? _____

In this type of structure, the C = O is derived from a *carboxyl* group and the N — H from an *amino* group.

The hydrogen bonds in the diagram in problem 388 exist between the H attached to a(n) _____ group and the oxygen attached to a(n) _____ group.

**hydrogen bond**

**amino; carboxyl**

**hydrogen bond**

Hydrogen bonding accounts for the pleated sheet structure of some proteins.

Some proteins have a helical structure, as shown in the following diagram.

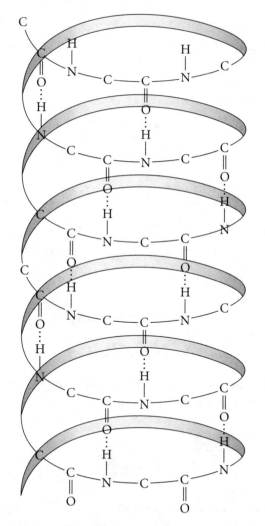

Helical structure of a protein

What type of bond holds the coil in its designated shape? _____

In both the pleated sheet and helical structures of proteins, the hydrogen bonds present exist between the H attached to a(n) _____ and the O that is part of a(n) _____ group.

amino; C = O from a carboxyl

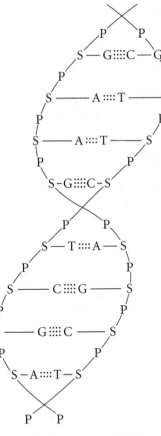

Consider the double helical shape of DNA in the following diagram.

Helical structure of DNA

hydrogen bond

What type of bond holds the two halves together? _____

2; hydrogen bonds

### 393

The DNA structure contains the following substances: adenine (A), guanine (G), cytosine (C), and thymine (T).

How many bonds are present between A and T? _____

What type of bonds are they? _____

3; hydrogen bonds

### 394

How many bonds are present between C and G? _____

What type of bonds are they? _____

weak

### 395

When the DNA molecule replicates, the two halves come apart, breaking the bonds holding them together.

Are hydrogen bonds strong or weak bonds? _____

easy

### 396

Should it be easy or difficult to "open up" the DNA chain? _____

# Isomers 12

Isomers are compounds that have the same atoms (same molecular formulas) but different three-dimensional arrangement of those atoms. Hence, isomers have different properties from one another.

Isomers can be classified as

a.  structural isomers

b.  geometric isomers

c.  enantiomers

## STRUCTURAL ISOMERS

### 397

Consider the following two compounds.

$$CH_3CH_2CH_2CH_3 \quad \text{and} \quad \overset{\displaystyle CH_3}{\underset{}{\overset{\displaystyle |}{CH_3CHCH_3}}}$$

a                                    b

What is the molecular formula for compound a? _____

What is the molecular formula for compound b? _____

$C_4H_{10}; C_4H_{10}$

no

Compounds a and b are structural isomers because they have the same molecular formula but different three-dimensional structures.

Would compounds a and b have identical properties? _____

C₂H₆O; C₂H₆O; yes; no

Consider the following two structures.

$$CH_3OCH_3 \qquad CH_3CH_2OH$$

Methyl ether        Ethanol

What is the molecular formula for methyl ether? _____

What is the molecular formula for ethanol? _____

Are methyl ether and ethanol structural isomers? _____

Do methyl ether and ethanol have the same properties? _____

## 400

Are the following pairs structural isomers?

a. _____

b. _____

c. _____

d. _____

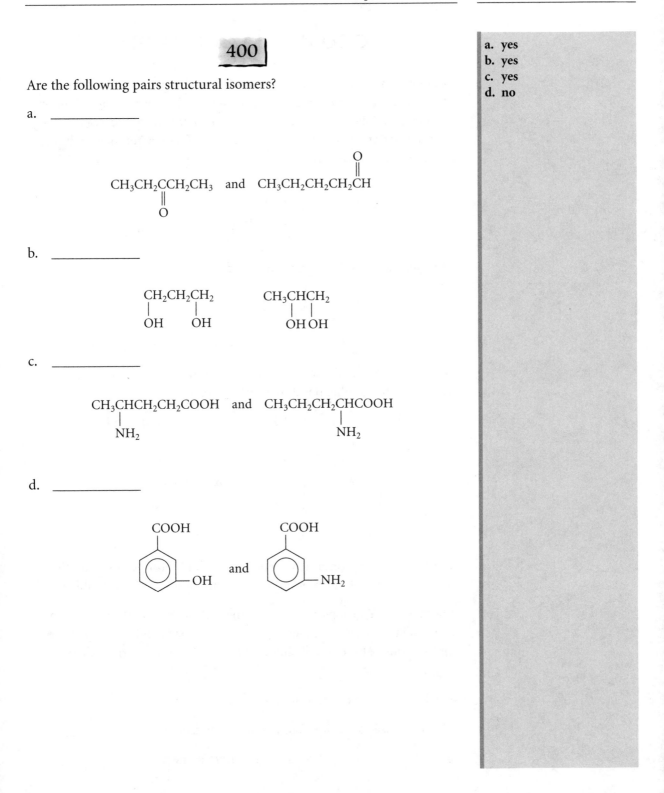

a. yes
b. yes
c. yes
d. no

# GEOMETRIC ISOMERS

Geometric isomers have the same molecular formula but different three-dimensional structures due to the presence of a double bond. A double bond is inflexible and hence does not allow the atoms to rotate freely. Such a restriction in rotation can account for a difference in biologic activity for those isomers.

**a; b**

Consider the following arrangement of carbon atoms

$$\begin{array}{cc} X & Y \\ | & | \\ -C & = C - -- \end{array}$$

where **X** and **Y** are the same or different atoms or groups of atoms attached to the double bond in a carbon chain.

Two possible arrangements of **X** and **Y** in such a compound are

Note that in such geometric isomers, the **X** and **Y** groups can be on the "same side" of the double bond or on "opposite sides" of the double bond.

If both **X** and **Y** groups are on the same side of the double bond, then the geometric isomer is called a *cis* isomer. If the **X** and **Y** groups are on opposite sides of the double bond, then such a compound is called a *trans* isomer.

Which of the above is an example of a *cis* isomer? _____

Which of the above is an example of a *trans* isomer? _____

Geometric isomers are also called *cis-trans* isomers.

## 402

a; b and c

Which of the following compounds is(are) a *cis* isomer? _____

Which is(are) a *trans* isomer? _____

a.

$$CH_3 - \overset{\overset{\displaystyle Cl}{|}}{C} = \overset{\overset{\displaystyle Cl}{|}}{C} - CH_2 - CH_3$$

b.

$$CH_3CH_2 - \overset{\overset{\displaystyle Cl}{|}}{C} = \underset{\underset{\displaystyle Cl}{|}}{C} - CH_2 - CH_2COOH$$

c.

$$CH_3 - \overset{\overset{\displaystyle CH_2CH_3}{|}}{C} = \underset{\underset{\displaystyle CH_2CH_3}{|}}{C} - CH_2 - CH_3$$

## 403

*cis*

Suppose that a carbon compound contains two double bonds such as

$$C - C = C - C - C = C - C$$

$\qquad\qquad$ **a.** $\qquad\qquad\qquad$ **b.**

If the double bond labeled **a** has substances attached on the same side of that double bond, would the compound be a *cis* or *trans* isomer?

_____

## 404

one where each double
bond would have a *trans*
arrangement; no

If the double bond labeled **b** also had a *cis* arrangement, then the isomer would be labeled a *cis-cis* isomer.

What type of structure would a *trans-trans* isomer be? _____

Would a pair of *cis-trans* isomers have the same properties?

_____

# ENANTIOMERS

Enantiomers are compounds with the same molecular formula but different three-dimensional structures that are nonsuperimposible mirror images of each other.

**no**

405

The mirror image of an object is the reflection of that object in a plane mirror.

If you hold your left hand in front of a plane mirror, you will see its mirror image, one that is identical to your right hand.

Left hand                    Mirror image of left hand

Is your left hand superimposible on your right hand? _____

406

Look at the structure for glyceraldehyde and its mirror image.

Glyceraldehyde

Mirror image of Glyceraldehyde

These same two structures can be represented as

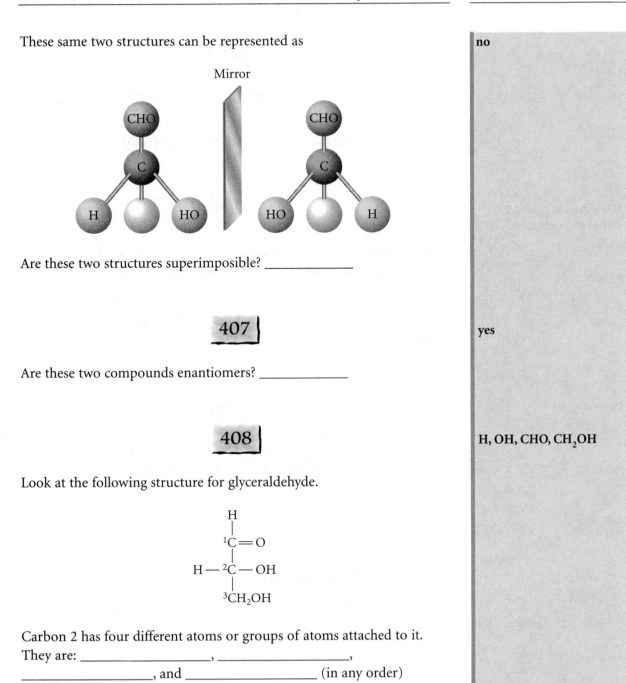

Are these two structures superimposible? _____

| 407 |

Are these two compounds enantiomers? _____

| 408 |

Look at the following structure for glyceraldehyde.

$$^1C=O \atop H-^2C-OH \atop ^3CH_2OH$$

Carbon 2 has four different atoms or groups of atoms attached to it.
They are: _____, _____,
_____, and _____ (in any order)

no; no

## 409

Does carbon 1 have four different atoms or groups attached to it? _____

Does carbon 3 have four different groups attached to it? _____

No, it does not have 4 different atoms attached to it.

## 410

A carbon with four different atoms or group of atoms attached to it is called an *asymmetric* carbon atom. Is carbon 1 in problem 408 asymmetric? _____ Why or why not? _____

Enantiomers (also called optical isomers) occur because of the presence of one or more asymmetric carbon atoms.

contains; yes

## 411

Glyceraldehyde _____ (contains/does not contain) an asymmetric carbon. Should glyceraldehyde have enantiomers? _____

no; no

## 412

Methane, $CH_4$, has the structure

$$H-\overset{\overset{\displaystyle H}{|}}{\underset{\underset{\displaystyle H}{|}}{C}}-H$$

Does methane have an asymmetric carbon? _____

Should methane have enantiomers? _____

413

The number of enantiomers (isomers) possible for a compound containing an asymmetric carbon atom is given by the formula

$$2^n$$

where $n$ is the number of asymmetric carbon atoms present.

Glyceraldehyde has 1 asymmetric carbon atom. How many isomers are possible for glyceraldehyde? _____

414

Look at the following structure.

$$
\begin{array}{c}
{}^{1}\text{CHO} \\
| \\
\text{H} - {}^{2}\text{C} - \text{OH} \\
| \\
\text{H} - {}^{3}\text{C} - \text{OH} \\
| \\
{}^{4}\text{CH}_2\text{OH}
\end{array}
$$

Carbon 1 has which groups attached to it? _____

415

Carbon 2 has which groups attached to it? _____

416

Carbon 3 has which groups attached to it? _____

---

$2^1$ or 2

$$
-\text{H}, =\text{O},\quad
\begin{array}{c}
\text{H} - \text{C} - \text{OH} \\
| \\
\text{H} - \text{C} - \text{OH} \\
| \\
\text{CH}_2\text{OH}
\end{array}
$$

$$
\text{CHO, H, OH,}\quad
\begin{array}{c}
\text{H} - \text{C} - \text{OH} \\
| \\
\text{CH}_2\text{OH}
\end{array}
$$

$$
\begin{array}{c}
\text{CHO} \\
| \\
\text{H} - \text{C} - \text{OH, H, OH, CH}_2\text{OH}
\end{array}
$$

CHO
|
H — C — OH, H, H, OH
|
H — C — OH

**2 and 3**

**2² or 4**

**4**

Carbon 4 has which groups attached to it? _____

Which of the above indicated carbons (1, 2, 3, or 4) is(are) asymmetric?

_____

The compounds indicated in problem 414 have how many optical isomers? _____

Glucose, $C_6H_{12}O_6$, has the following structure.

$$
\begin{array}{c}
H \\
| \\
C=O \\
| \\
H - C - OH \\
| \\
HO - C - H \\
| \\
H - C - OH \\
| \\
H - C - OH \\
| \\
CH_2OH
\end{array}
$$

Glucose has how many asymmetric carbons? _____

**421**

Glucose has how many isomers? _____

**422**

Glucose contains which functional group(s)? _____

$2^4$ or 16

carbonyl and hydroxyl

## PART III

Biochemistry

# Carbohydrates 13

Carbohydrates contain the elements carbon, hydrogen, and oxygen, and only those elements.* With few exceptions, a compound containing elements other than these three is not a carbohydrate.

The word *hydrate* refers to water. Many carbohydrates contain hydrogen and oxygen in the same ratio as in water, 2:1.

Which of the following compounds are carbohydrates?

___ $C_6H_6$          ___ $C_{12}H_{22}O_{11}$     ___ $C_5H_{10}O_5$
___ $C_6H_{12}O_6NS$     ___ $C_2H_6O$

$C_{12}H_{22}O_{11}$ and $C_5H_{10}O_5$

### 424

a.  In the list of compounds in problem 423, why is $C_6H_{12}O_6NS$ not a carbohydrate? _____
_____

b.  Why are $C_6H_6$ and $C_2H_6O$ not carbohydrates? _____
_____
_____

a. $C_6H_{12}O_6NS$ contains nitrogen and sulfur, and carbohydrates contain only C, H, and O

b. $C_6H_6$ contains no oxygen, and $C_2H_6O$ has the wrong ratio of H to O

_____

* An exception to this generalization is *chitin*, a structural polysaccharide (carbohydrate) used by arthropods to build their exoskeletons and also used by some fungi. Chitin is composed of many glucose molecules $(C_6H_{12}O_6)$ modified with a nitrogen-containing compound.

**a. one**
**b. one**

### 425

Carbohydrates are divided into three types. One type of carbohydrate is the *monosaccharide.*

a.   The word *saccharide* means *simple sugar.* The prefix *mono* means

_____.

b.   Therefore, monosaccharides are simple sugars, each containing only _____ simple sugar.

**all of them**

### 426

Monosaccharides and disaccharides have names ending in *ose.* Which of the following are monosaccharides or disaccharides?
___ glucose      ___ fructose      ___ galactose
___ arabinose   ___ erythrose

**5**

### 427

Monosaccharides are named according to the number of carbon atoms they contain. A pentose contains _____ carbon atoms.

An example of a pentose is ribose, $C_5H_{10}O_5$. The structure of ribose is

In this type of structure, the ring, consisting of carbon atoms at each corner and an oxygen atom where indicated, is considered to be in a plane perpendicular to the paper. The heavy line in the ring indicates the section closest to you; the light lines, the part farther from you. The —H and —OH (hydroxyl) groups are above and below the plane as indicated.

Ribose occurs in ribonucleic acid, RNA, which will be discussed later.

### 428

a. one less oxygen
b. one less oxygen

Compare the structures of ribose, $C_5H_{10}O_5$, and deoxyribose, $C_5H_{10}O_4$.

Ribose                    Deoxyribose

a.   The prefix *deoxy* means _____.

b.   Deoxyribose contains _____ than does ribose.

### 429

DNA

Ribose occurs in ribonucleic acid, RNA. Deoxyribose occurs in deoxyribonucleic acid, _____.

a. 5
b. 6

## 430

a.  A pentose is a monosaccharide containing _____ carbon atoms.

b.  A hexose is a monosaccharide containing _____ carbon atoms.

A primary hexose in the body is glucose, $C_6H_{12}O_6$. The oxidation of glucose supplies much of the energy the body needs. The structure of glucose is

ose

## 431

The second type of carbohydrate is the *disaccharide*. Both monosaccharides and disaccharides have names ending in _____.

## 432

The prefix *di* means *two*. Disaccharides are formed by the combination of two monosaccharides, with water also being produced. An enzyme is required for this reaction.

monosaccharide + monosaccharide $\xrightarrow{\text{enzyme}}$ disaccharide + water

Conversely, disaccharides react with water to produce two monosaccharides.

disaccharide + water $\xrightarrow{\text{enzyme}}$ monosaccharide + _____

These reactions may be written as shown on the following page.

The bond holding the two halves of the disaccharide together is called a *glycosidic bond*, or a *glycosidic linkage*.

**monosaccharide**

## 433

When a disaccharide reacts with water, two monosaccharides are produced. Such a reaction is called *hydrolysis*, which is defined as the breaking apart of a molecule by reaction with water.

Digestion involves hydrolysis. During digestion, molecules are:

__ broken down into smaller molecules
__ built up into larger molecules

**broken down into smaller molecules**

**monosaccharides**

434

Disaccharides that have undergone hydrolysis yield two _____.

**one; two**

435

Monosaccharides are made up of _____ simple sugar(s).
Disaccharides are made up of _____ simple sugar(s).

---

### 436

The hydrolysis of a disaccharide may be written as:

$$C_{12}H_{22}O_{11} + H_2O \xrightarrow{\text{enzyme}} C_6H_{12}O_6 + C_6H_{12}O_6$$

Disaccharide + water            Monosaccharide + monosaccharide

Sucrose, $C_{12}H_{22}O_{11}$, is a disaccharide. The hydrolysis of sucrose yields the two _____, glucose and fructose.

 monosaccharides

### 437

There are three common disaccharides: *sucrose*, *maltose*, and *lactose*. These disaccharides all have the same molecular formula, $C_{12}H_{22}O_{11}$. Lactose, also called milk sugar, is found in milk. Maltose, or malt sugar, is found in sprouting grain. Sucrose, or cane sugar, is found in _____.

 sugarcane

### 438

The third type of carbohydrate is the *polysaccharide*. The prefix *poly* means *many*, so polysaccharides, on hydrolysis, yield many _____.

 monosaccharides

### 439

Recall that hydrolysis means reaction with _____ to produce _____ (smaller/larger) molecules.

 water; smaller

$C_{12}H_{22}O_{11}$

## 440

Starch is a polysaccharide. Its molecular formula is $(C_6H_{10}O_5)_n$, where $n$ is some large number. Upon complete hydrolysis, starch yields many monosaccharides:

$$(C_6H_{10}O_5)_n \ + \ nH_2O \ \xrightarrow{\text{enzyme}} \ nC_6H_{12}O_6$$

Starch,
a polysaccharide     Monosaccharides

Which of the following formulas represents a disaccharide?

___ $C_5H_{10}O_5$          ___ $C_{12}H_{22}O_{11}$
___ $(C_6H_{10}O_5)_n$       ___ $C_6H_{12}O_6$

$(C_6H_{10}O_5)_n$; $C_5H_{10}O_5$
and $C_6H_{12}O_6$

## 441

Which of the formulas listed in problem 440 represents a polysaccharide? _____

Which are monosaccharides? _____

a. glucose, fructose,
   galactose
b. lactose, maltose,
   sucrose
c. starch, cellulose,
   glycogen

## 442

Other examples of polysaccharides are *cellulose*, found in plants, and *glycogen*, found in animals. Plants use the polysaccharide cellulose primarily for support. Animals store carbohydrate in the form of a polysaccharide, glycogen.

a.  Name three monosaccharides. _____

b.  Name three common disaccharides. _____

c.  Name three polysaccharides. _____

Disaccharides are held together by a glycosidic bond. Polysaccharides are held together by _____ bonds.

many glycosidic

---

### 444

a. Hydrolysis of a disaccharide yields _____.

b. Hydrolysis of a polysaccharide yields _____.

a. two monosaccharides
b. many mono-
saccharides

# PROPERTIES
# OF CARBOHYDRATES

*Monosaccharides* are crystalline compounds, soluble in water and sweet to the taste. They need no digestion to be absorbed into the bloodstream.

*Disaccharides* are crystalline, water soluble, and sweet to the taste, and must be hydrolyzed (digested) to monosaccharides before they can be absorbed and used by the body for energy.

*Polysaccharides* are not water soluble and are not crystalline. They form colloidal dispersions instead of solutions. They are not sweet and must be hydrolyzed before being absorbed. (Cellulose is not hydrolyzed by enzymes in the human gastrointestinal tract.)

a. mono- and
   disaccharides
b. polysaccharides
c. mono- and
   disaccharides
d. polysaccharides
e. mono- and
   disaccharides
f. polysaccharides
g. monosaccharides

## 445

a.  Which type(s) of carbohydrate is(are) soluble in water? _____

b.  Which type(s) of carbohydrate form(s) a colloidal suspension in water? _____

c.  Which type(s) of carbohydrate is(are) crystalline? _____

d.  Which is(are) not crystalline? _____

e.  Which is(are) sweet? _____

f.  Which is(are) not sweet? _____

g.  Which require(s) no digestion? _____

# Lipids 14

Lipids include such substances as fats, phospholipids, and steroids. Lipids are insoluble in water. Lipids contain the same elements as carbohydrates, but in lipids the proportion of oxygen is much lower than in carbohydrates.

**446**

Lipids contain the elements _____, _____, and _____.

**carbon; hydrogen; oxygen**

## FATS

**447**

Fats are formed by the reaction of an organic acid with an alcohol. In fats, the organic acid is called a *fatty acid* and the alcohol is *glycerol* (sometimes called glycerin). The reaction is

$$\text{fatty acids} + \text{glycerol} \xrightarrow{\text{enzymes}} \text{fat} + \text{water}$$

The reverse reaction is the hydrolysis of a fat.

Thus, the hydrolysis of a fat yields _____ + _____.

**fatty acids; glycerol**

**448**

Fatty acids and glycerol are products of the hydrolysis of a _____.

**fat**

fatty acid; glycerol

## 449

The chemical equation for the hydrolysis of a fat may be written as

Substance A is a fat.

Substance B is a _____.

Substance C is called _____.

a.  contains carboxyl
    (—COOH) groups
b.  contains hydroxyl
    (—OH) groups

## 450

a.  How do you know that substance B is an acid? _____

b.  How do you know that substance C is an alcohol? _____

a.  many mono-
    saccharides
b.  two monosaccharides
c.  fatty acids and
    glycerol

## 451

a.  The hydrolysis of a polysaccharide yields _____.

b.  The hydrolysis of a disaccharide yields _____.

c.  The hydrolysis of a fat yields _____.

Fatty acids that contain double bonds are called *unsaturated.*
Unsaturated fatty acids tend to be more liquid than saturated fatty
acids. A liquid fat is called an *oil.*

Increasing the number of double bonds between carbon atoms in
fatty acids of a fat molecule should make the fat more _____
(solid/liquid). Decreasing the number of double bonds in a fat will
make the fat more _____ (solid/liquid).

Which are unsaturated, fats or oils? _____

Unsaturated compounds contain double bonds. What kind of
compounds would be called polyunsaturated? _____

455

Three fatty acids, butyric ($C_3H_7COOH$), palmitic ($C_{15}H_{31}COOH$),
and stearic ($C_{17}H_{35}COOH$) acid, contain only single bonds. When
they are present in a lipid, will that lipid tend to become solid or
liquid? _____

Arachidonic acid is unsaturated, which means that it contains
_____ (only single bonds/some double bonds) in its structure.

---

liquid; solid

oils

those containing many
double bonds

solid

some double bonds

**an oil; the fatty acid is unsaturated**

A lipid that undergoes hydrolysis yields arachidonic acid and glycerol. Was the lipid a fat or an oil? _____

How do you know? _____

**nonpolar**

Lipids are insoluble in water. They are soluble in organic solvents such as gasoline, ether, carbon tetrachloride, and benzene. Organic solvents are nonpolar. Recall that, in general, polar compounds dissolve in polar liquids and nonpolar compounds dissolve in nonpolar liquids. Are lipids polar or nonpolar? _____

**oleic acid is a *cis*-fatty acid; elaidic acid is a *trans*-fatty acid**

Consider the following two fatty acids, oleic acid and elaidic acid.
Which is a *cis*-fatty acid? _____
Which is a *trans*-fatty acid? _____
(If you are not sure, see problem 401.)

Oleic acid

Elaidic acid

# PHOSPHOLIPIDS

## 460

The hydrolysis of a phospholipid may be written as

phospholipid $\xrightarrow[\text{enzymes}]{\text{hydrolysis}}$ fatty acids + glycerol
+ phosphoric acid + an N-compound

The hydrolysis of a fat yields _____ + _____.

## 461

The N-compounds in phospholipids are a certain group of nitrogen-containing compounds. Among them are serine, choline, inositol, and ethanolamine. Which is present determines the type of phospholipid, as will be discussed in the following problems.

The hydrolysis of a phospholipid yields _____
_____.

## 462

a. Lipids contain which elements? _____

b. Phospholipids contain which elements? _____

## 463

Lecithin is a phospholipid in which the N-compound is *choline*. Lecithin is a main component of cell membranes and is involved in the transportation of fats in the body. The hydrolysis of lecithin yields fatty acids, glycerol, phosphoric acid, and _____.

**fatty acids; glycerol;
phosphoric acid;
ethanolamine**

464

Cephalin is a phospholipid involved in the blood-clotting process. In cephalin, the N-compound is *ethanolamine*.

cephalin  _____, _____, _____, and _____.

# STEROIDS

**a. insoluble
b. insoluble
c. soluble**

465

Another group, called steroids, is classified along with lipids because of similar solubility characteristics.

a.   Lipids are _____ (soluble/insoluble) in water.

b.   Steroids are _____ (soluble/insoluble) in water.

c.   Steroids are _____ (soluble/insoluble) in organic liquids.

**a. no
b. no**

466

Steroids are high-molecular-weight tetracyclic (4-ring) compounds whose structure is similar to

a.   Do steroids contain fatty acids? _____

b.   Do steroids contain glycerol? _____

Examples of steroids in the body are cholesterol, bile salts, sex hormones, and hormones of the adrenal cortex.

a.   Does cholesterol have a steroid structure? _____

b.   The ending *ol* in cholesterol indicates that it is what type of compound? _____

It contains which functional group? _____

# Proteins 15

## FORMATION AND HYDROLYSIS

carbon, hydrogen, oxygen, nitrogen (in any order)

### 468

Proteins contain the same elements as do carbohydrates and fats, except that proteins always contain one additional element: nitrogen.

All proteins contain the four elements: _____.

> carbon, hydrogen, oxygen, nitrogen (in any order)

### 469

Some proteins also contain additional elements, such as sulfur, phosphorus, and iron. The hydrolysis of proteins yields amino acids. Likewise, the combination of amino acids yields proteins.

$$\text{protein} \underset{\text{combination}}{\overset{\text{hydrolysis}}{\rightleftharpoons}} \text{amino acids}$$

Amino acids contain what functional groups? _____

> amino and carboxyl

— COOH or carboxyl

### 470

The structure of an amino acid can be represented as

where "R" represents either a hydrogen or carbon-hydrogen groups.

Amino acids (and also proteins) can act as buffers because the $-NH_2$ group acts as a $H^+$ acceptor and the _____ group acts as a $H^+$ donor.

a. 3
b. many (or more than 3)

### 471

When two amino acids combine, the product is called a *dipeptide*.

a.  A *tripeptide* would be formed when _____ amino acids combine.

b.  A *polypeptide* would be formed when _____ amino acids combine.

amino acids

### 472

Note the formulas for alanine and glycine:

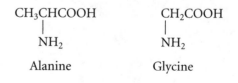

Both would be classified as what type of compound? _____

## 473

When these two amino acids react, the following reactions are possible.

Alanine       Glycine       Alanyl-glycine (ala-gly)

Glycine       Alanine       Glycyl-alanine (gly-ala)

Note that in both of these equations, the reaction is between the OH of a(n) _____ group and the H of a(n) _____ group.

## 474

When glycine and alanine react, what type of compound is formed?

_____

## 475

a. The abbreviation *ala* represents _____.

b. The abbreviation *ala-gly* represents _____.

a tripeptide

If the amino acids glycine, alanine, and valine were combined in one compound, what type would it be? _____

ala-gly-val

What would be the abbreviation for a tripeptide containing alanine, glycine, and valine, in that order? _____

a tripeptide containing glycine, valine, and alanine, combined in that order

**478**

What would the abbreviation *gly-val-ala* represent? _____
_____

many; peptide

**479**

A dipeptide consists of two amino acids held together by a *peptide bond.* A tripeptide consists of three amino acids held together by peptide bonds. A polypeptide consists of _____ amino acids held together by _____ bonds.

carboxyl; amino

In a peptide bond (see problem 473), the bond occurs between a(n) _____ group of one amino acid and a(n) _____ group of another amino acid.

# STRUCTURE

The *primary structure* of a protein refers to the linear sequence of amino acids in that protein. An example of a primary structure is

Amino acid          Amino acid          Amino acid

### 481

What type of bond holds the amino acids together in the primary structure? _____

Note that all proteins have an amino ($-NH_2$) terminus and a carboxyl ($-COOH$) terminus.

peptide

The *secondary structure* of a protein refers to the regular recurring arrangement of the amino acid chain (its primary structure). One such arrangement, called the $\alpha$-helix, occurs when the amino acid chain forms a spiral or coil, as shown here.

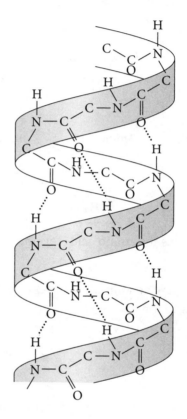

**hydrogen**

482

The coil consists of loops of the linear arrangement of the amino acids. These loops are held together by bonds between the O of the C=O of one amino acid and the H of the NH of another amino acid.

Such a bond is called a(n) _____ bond (refer to problems 388–392 if you don't know).

hydrogen

Another type of secondary structure, called the β-pleated sheet, consists of parallel strands of polypeptides, as shown in the following figure.

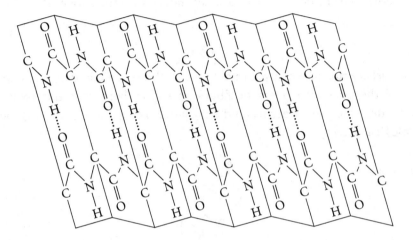

What type of bonds holds these sheets together? _____

a. The linear sequence of amino acids in a protein is called the _____ structure of that protein.

b. The regular recurring arrangement of the amino acid chains is called the _____ structure of a protein.

**hydrogen**

**a. primary**
**b. secondary**

**a.** *α*-helix; *β*-pleated
   sheet
**b.** hydrogen

a.   Two types of secondary structures are possible. They are the
     _____ and the _____.

b.   What type of bond holds these secondary structures together?
     _____

The *tertiary structure* of a protein refers to the specific folding and bending of the coils into layers or fibers, as shown here. It is this tertiary structure that gives proteins (and enzymes, which are also proteins) their biological activity.

Some proteins have a *quaternary structure* that occurs when several protein units, each with its own primary, secondary, and tertiary structure, combine to form a more complex unit, as shown in figure e. An example of a protein with a quaternary structure is hemoglobin.

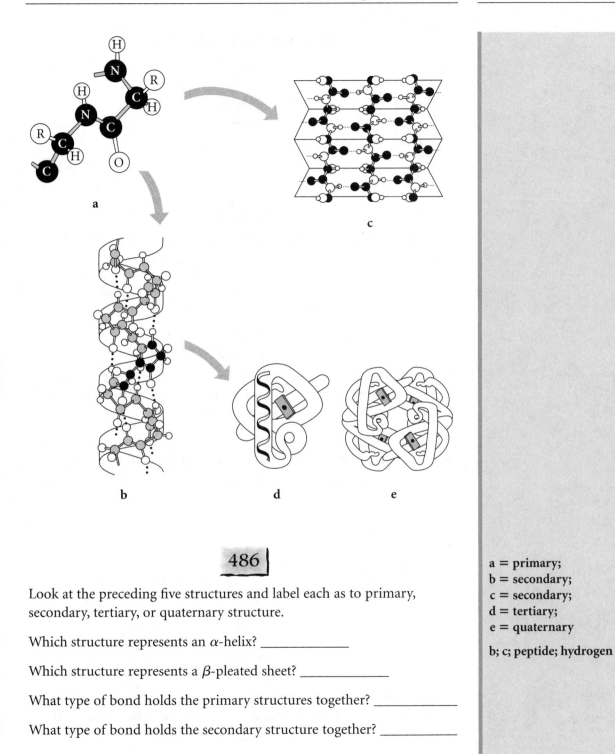

a

c

b

d

e

486

Look at the preceding five structures and label each as to primary, secondary, tertiary, or quaternary structure.

Which structure represents an α-helix? _____

Which structure represents a β-pleated sheet? _____

What type of bond holds the primary structures together? _____

What type of bond holds the secondary structure together? _____

a = primary;
b = secondary;
c = secondary;
d = tertiary;
e = quaternary

b; c; peptide; hydrogen

# DENATURATION

Hydrogen bonds are _____ (strong/weak) bonds.

**488**

Therefore, hydrogen bonds are easily broken. If the hydrogen bonds in a protein are broken, can the protein easily maintain its structural shape? _____

**489**

If the hydrogen bonds in a protein are broken, then the shape of the protein can change, making it incapable of performing its physiologic function. The protein is said to be *denatured*.

The activity of many proteins (and enzymes) depends on the presence of an active site that "fits" into a specific substrate. In figure **a**, the active site is shown in the center of the structure.

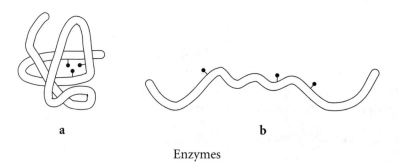

a                    b

Enzymes

If the protein is denatured, then the parts of the active site are no longer in close proximity (figure **b**), can no longer "fit" into the substrate, and therefore _____ (can/cannot) react.

If the hydrogen bonds in a protein can easily re-form, then the denaturation is said to be *reversible*. If the hydrogen bonds cannot re-form, then the process is termed *irreversible denaturation*.

a. In the denaturation of protein, _____ bonds are broken.

b. If the broken hydrogen bonds can easily re-form, then the process is called _____ denaturation.

c. If the broken hydrogen bonds cannot re-form, then the process is called _____.

Factors that can denature proteins include strong acids, heat, alcohol, and salts of heavy metals.

a. hydrogen
b. reversible
c. irreversible denaturation

## ADENOSINE TRIPHOSPHATE

### 491

Nucleotides are formed by the reaction of a pentose (labeled Ⓢ for sugar), phosphoric acid (labeled Ⓟ), and a nitrogen-containing base (labeled Ⓝ), as shown:

One example of a nucleotide is adenosine monophosphate, AMP, which is formed by the reaction of adenine (a nitrogen-containing base), ribose (a pentose), and phosphoric acid. From the name *adenosine monophosphate*, how many phosphoric acid molecules are present?

_____

### 492

The structure of adenosine monophosphate, AMP, is shown in the following figure. It can be abbreviated as adenosine-P.

**ribose; adenine**

AMP

AMP contains which pentose? _____

Which nitrogen-containing base? _____

3

### 493

If a second molecule of phosphoric acid is bonded to the first phosphoric acid group, then adenosine diphosphate, ADP, is formed. In adenosine triphosphate, ATP, how many phosphoric acid groups are present? _____

energy

### 494

ATP provides a form of chemical energy that is usable by all body cells. The structure of ATP may be abbreviated as

$$\text{adenosine} - \text{P} \sim \text{P} \sim \text{P}$$

Note that ATP contains two high-energy phosphate bonds indicated as $\sim$.

In the reaction

$$\text{ATP} \longrightarrow \text{ADP} + \text{P} + \text{?} \qquad \text{or}$$

$$\text{adenosine} - \text{P} \sim \text{P} \sim \text{P} \longrightarrow \text{adenosine} - \text{P} \sim \text{P} + \text{P} + \text{?}$$

what is produced other than ADP and P? _____

## 495

When ADP is converted into AMP, is energy released? _____

yes

## 496

To convert AMP back to ATP, what must be added? _____

energy and two phosphoric acid groups

## 497

Much of the energy required to form ATP comes from the metabolism of glucose.

a.   AMP contains _____ phosphoric acid group(s) and _____ high-energy phosphate bond(s).

b.   ADP contains _____ phosphoric acid group(s) and _____ high-energy phosphate bond(s).

c.   ATP contains _____ phosphoric acid group(s) and _____ high-energy phosphate bond(s).

a.  1; 0
b.  2; 1
c.  3; 2

## 498

When two nucleotides are joined together, a dinucleotide is formed. One such dinucleotide is nicotinamide adenine dinucleotide, NAD.

a.   Both nucleotides that make up NAD must contain which pentose?

_____

b.   One nucleotide contains nicotinamide as the nitrogen-containing base (N-base); the other contains _____ as the nitrogen-containing base.

a.  ribose
b.  adenine

**flavin, adenine**

Which nitrogen-containing bases are present in flavin adenine dinucleotide, FAD? _____

Both dinucleotides, FAD and NAD, are involved in oxidation-reduction reactions in cells.

# NUCLEIC ACIDS: DNA AND RNA

Nucleic acids are polymers of nucleotides. Consider the following structure of deoxyribonucleic acid, DNA, a nucleic acid. Note that each boxed segment is a nucleotide.

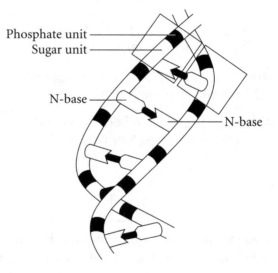

DNA consists of a double coil with the opposite sides held together by hydrogen bonds. The backbones of the coils are alternating sugar and phosphate groups, with the hydrogen bonding between the nitrogen-containing bases holding the sides together.

### 500

In deoxyribonucleic acid, which pentose is a part of the coil backbone?

_____

**deoxyribose**

### 501

In DNA, there are four N-bases: adenine, guanine, thymine, and cytosine. They are abbreviated by the first letter (capitalized) of their name.

Adenine is abbreviated as _____.

Guanine is abbreviated as _____.

Thymine is abbreviated as _____.

Cytosine is abbreviated as _____.

**A; G; T; C**

### 502

Which N-base is represented by each of the following abbreviations?

T _____

A _____

C _____

G _____

**thymine; adenine; cytosine; guanine**

2; 3

**503**

In DNA, adenine is always hydrogen-bonded to thymine, and cytosine is always hydrogen-bonded to guanine. Thus, the structure of DNA may be represented as

How many hydrogen bonds hold A and T together? _____

How many hydrogen bonds hold C and G together? _____

adenine, A; thymine, T; guanine, G; and cytosine, C

The four N-bases in DNA are _____

_____.

In DNA, adenine is always hydrogen-bonded to _____, and guanine is always bonded to _____.

thymine; cytosine

Complete the structure of the following figure, indicating missing N-bases and hydrogen bonds.

A second nucleic acid is ribonucleic acid, RNA. In RNA, the pentose is ribose; in DNA, the pentose is _____.

deoxyribose

In RNA, A is always bonded to uracil, U. In DNA, A is always bonded to _____.

T

thymine; cytosine

a. G; 3
b. G; 3
c. T; 2
d. U; 2
e. deoxyribose
f. ribose

## 509

a.   In DNA, C bonds to _____ with _____
     (how many?) hydrogen bonds.

b.   In RNA, C bonds to _____ with _____
     (how many?) hydrogen bonds.

c.   In DNA, A bonds to _____ with _____
     (how many?) hydrogen bonds.

d.   In RNA, A bonds to _____ with _____
     (how many?) hydrogen bonds.

e.   In DNA, the sugar is _____.

f.   In RNA, the sugar is _____.

# REPLICATION OF DNA

## 510

Consider the following illustration in which a DNA segment has been
straightened out and the S's (sugars) and P's (phosphates) are understood and not printed.

```
    T   C   A   A   G
    ‖   ‖‖  ‖   ‖   ‖‖
    A   G   T   T   C
```

If the DNA strands are separated, then each strand can act as a template
to attract additional N-bases.

                            T   C   A   A   G

Original
strand
opened up
                            A   G   T   T   C

Let us consider the T and A at the far left of the open strand. The T will attract an A, and the A will attract a T to begin the formation of a new backbone of a DNA molecule, as follows.

Now complete the balance of the strands for each DNA molecule, again showing N-bases and hydrogen bonding.

Are the two new DNA strands the same as the original one?

_____

yes

# TRANSCRIPTION OF INFORMATION (PROTEIN SYNTHESIS)

### 512

During transcription, DNA "unzips" and only part of one segment acts as a template for the formation of messenger RNA (mRNA).

In DNA, the partner of adenine (A) is _____.

In mRNA, the partner of adenine (A) is uracil, or _____.

T, or thymine; U

2; U; 2

In DNA, A is bonded to T with how many hydrogen bonds?
_____

Likewise in mRNA, A is bonded to _____ with how many hydrogen bonds? _____

deoxyribose; ribose

### 513

In DNA, the sugar is _____.

In mRNA, the sugar is _____.

$$
\begin{array}{cccccc}
\text{C} & \text{A} & \text{G} & \text{T} & \text{A} & \text{C} \\
\vdots & \vdots & \vdots & \vdots & \vdots & \vdots \\
\text{G} & \text{U} & \text{C} & \text{A} & \text{U} & \text{G}
\end{array}
$$

### 514

Complete the following diagram indicating the partners and bonds needed for the formation of mRNA from the indicated DNA segment.

Segment of DNA    ——————————————
                  C   A   G   T   A   C

mRNA    ————————————————————

mRNA acts as a template for the transmission of genetic information from the cell's genes to the site of protein synthesis.

In the cytoplasm, another type of RNA is present. It is called transfer RNA or tRNA. The function of tRNA is to direct the production of the protein called for by mRNA.

tRNA has three N-bases at one end and an amino acid at the other end. It may be indicated by the following structure.

Amino acid

N-bases
or codons

As mRNA travels across a ribosome, the first group of three
N-containing bases picks up the corresponding tRNA. As the mRNA
continues through the ribosome, the second group of three N-bases
picks up its corresponding tRNA. The amino acid attached to the second tRNA bonds to the one attached to the first tRNA, and then the
third one bonds to the second, and so on until a chain of amino acids (a
protein) is formed, as in the following figure.

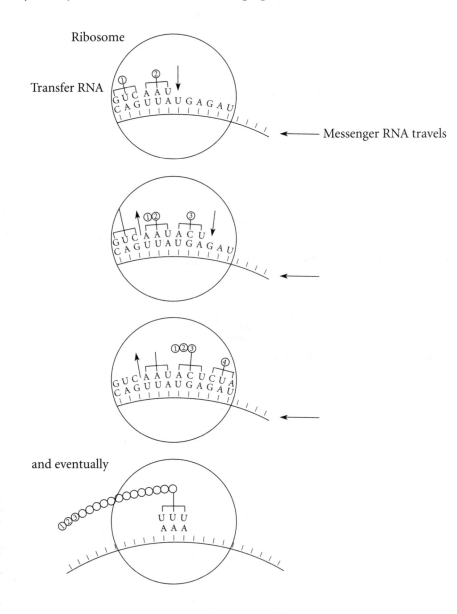

3

tRNA binds to mRNA at three consecutive sites, as in the following figure.

mRNA

| A | C | C | U | G | A | G | C | U |

U  G  G     A  C  U     C  G  A

Amino acid   Amino acid   Amino acid

When tRNA binds to mRNA, how many bonding sites are there?

_____

These three binding sites are called the *triplet code* or *codon*.

**UGG**

In the diagram in problem 515, the beginning triplet code ACC corresponds to which code in tRNA? _____

Indicate the tRNAs and binding needed for the following segment of mRNA.

mRNA

| A | C | G | C | U | U | G | C | A | U | C | C |

mRNA

A  C  G  C  U  U

tRNA    U  G  C   G  A  A

Amino acid  Amino acid

mRNA

G  C  A  U  C  C

tRNA    C  G  U   A  G  G

Amino acid  Amino acid

# THE GENETIC CODE

Each tRNA is specific for a certain amino acid. The following chart indicates the amino acids corresponding to the triplet code in mRNA. Note that there is some redundancy; some amino acids are coded for more than one codon.

**Messenger RNA**

| First Letter | Second Letter | Third Letter | | | |
|---|---|---|---|---|---|
| | | A | C | G | U |
| A | A | lys | asn | lys | asn |
| | C | thr | thr | thr | thr |
| | G | arg | ser | arg | ser |
| | U | ile | ile | met* | ile |
| C | A | gln | his | gln | his |
| | C | pro | pro | pro | pro |
| | G | arg | arg | arg | arg |
| | U | leu | leu | leu | leu |
| G | A | glu | asp | glu | asp |
| | C | ala | ala | ala | ala |
| | G | gly | gly | gly | gly |
| | U | val | val | val | val |
| U | A | ** | tyr | ** | tyr |
| | C | ser | ser | ser | ser |
| | G | ** | cys | trp | cys |
| | U | leu | phe | leu | phe |

*Chain initiator
**Chain terminator

518

thr

Look at the answer in problem 517. The first tRNA bonds to the coded group ACG in mRNA. According to the preceding table, ACG in the mRNA corresponds to which amino acid? _____

leu; ala; ser

## 519

The second tRNA in problem 517 corresponds to which amino acids? _____ The third tRNA? _____ The fourth? _____

The DNA segment above calls for four amino acids that are bonded in this order:

thr-leu-ala-ser

lys-arg-glu-ser

## 520

Indicate the amino acid chain designated by the following strand of mRNA.

A　A　A　C　G　G　G　A　G　U　C　C

_____

AUG; UAA, UAG, UGA

## 521

Which coded group(s) indicate the start of an amino acid chain? _____ The termination of the chain? _____

ACC GCU UUU;
thr-ala-phe

## 522

Look at the following segment of DNA.

T　G　G　C　G　A　A　A　A

What is the sequence of mRNA for this segment: _____

What is the sequence of amino acids indicated? _____

# MUTATIONS

### 523

**thr-ala-phe; thr-ala-leu**

Suppose that somehow one of the N-bases in the DNA in problem 522 was miscopied. Two such examples follow. In the first example, the sixth base is copied as a G instead of a U. In the second example, the last base is copied as an A instead of a U.

mRNA first miscopy: ACC GCG UUU

mRNA second miscopy: ACC GCU UUA

What sequence of amino acids would be indicated by the first miscopy? _____ By the second miscopy? _____

### 524

**no; maybe**

A change in copying DNA is called a mutation. Some mutations are not harmful; others might be. Would the change in the first miscopy cause a harmful mutation? _____ In the second miscopy? _____

GAA or GAG

## 525

The following diagram indicates the sequence of the 146 amino acids present in hemoglobin.

```
     1                5                      10                    15
   val - his - leu - thr - pro — glu — glu - lys - ser - ala - val - thr - ala - leu - try - gly - lys
                                                                                                        val
          30                    25                    20
    val - val - leu - leu - arg - gly - leu - ala - glu - gly - gly - val - glu - asp - val - asn
  35
  tyr
     pro - try - thr - gln - arg - phe - phe - glu - ser - phe - gly - asp - leu - ser - thr - pro
          40                    45                    50
                                                                                                        asp
          65                    60                    55
         leu - val - lys - lys - gly - his - ala - lys - val - lys - pro - asn - gly - met - val - ala
  gly
     70                    75                    80                    85
     ala - phe - ser - asp - gly - leu - ala - his - leu - asp - asn - leu - lys - gly - thr - phe
                                                                                                        ala
          100                   95                    90
         asn - glu - pro - asp - val - his - leu - lys - asp - cys - his - leu - glu - ser - leu - thr
  phe
     105                   110                   115
     arg - leu - leu - gly - asn - val - leu - val - cys - val - leu - ala - his - his - phe - gly
                                                                                                        120
                                                                                                        lys
          135                   130                   125
         val - gly - ala - val - val - lys - gln - tyr - ala - ala - gln - val - pro - pro - thr - phe - glu
  ala
     140                   146
     asn - ala - leu - ala - his - lys - tyr - his
```

Note that the sixth amino acid is glu. What is(are) the coded group(s) for glu? _____

GUA or GUG

## 526

Suppose that the middle section of either of these code groups is changed from A to U. Then the new coded groups would be either _____ or _____.

What amino acid(s) would this new coded group designate?

_____

val

528 |

Would this new protein of 146 amino acids still be hemoglobin?

_____

no

The new protein, with only one of 146 amino acids changed, is called hemoglobin-S and causes the genetic disease sickle-cell anemia.

# Enzymes 17

Enzymes are proteins that act as biological catalysts. A catalyst increases the rate of a reaction but is not changed itself.

Pepsin is an enzyme found in gastric juice. It increases the rate of reaction for the digestion of protein in the stomach. Note that pepsin is an enzyme for a particular substance: protein. In general, enzymes are highly specific; catalysts are nonspecific.

### 529

Enzymes are _____ (specific/nonspecific) catalysts.

specific

### 530

a. What effect does an enzyme have on the rate of a reaction?
_____

b. Is an enzyme changed by the reaction? _____

a. it increases it
b. no

# NAMES OF ENZYMES

**a. carbohydrates**
**b. lipids**

Most enzymes have names ending in *ase*. Some enzymes are named for the type of reaction they catalyze. *Hydrolases* are enzymes that catalyze hydrolysis reactions. *Oxidases* are enzymes that catalyze oxidation reactions.

a.  Carbohydrases are enzymes that catalyze the hydrolysis of _____.

b.  Lipases are enzymes that catalyze the hydrolysis of _____.

**a. lactose**
**b. maltase**

Other enzymes are named according to the substrate upon which they act. *Sucrase* is an enzyme that catalyzes the hydrolysis of sucrose.

a.  Lactase is an enzyme that catalyzes the hydrolysis of _____.

b.  Which enzyme catalyzes the hydrolysis of maltose? _____

**acceptor**

Another class of enzymes is the *transferase,* which is involved in the transfer of a specific group from a compound called a *donor* to another compound called an *acceptor.*

*Aminotransferase* is involved in the transfer of an amino group from a donor compound to another compound called a(n) _____.

**the transfer of a phospho or phosphate group from a donor to an acceptor**

What is the function of a *phosphotransferase?* _____

*Methyltransferases* are involved in the transfer of a _____ group from a _____ compound to a(n) _____ compound.

Methyltransferases are involved in turning genes "on" or "off" and will be discussed in Chapter 18.

**methyl, donor, acceptor**

# COENZYMES

Some enzymes are proteins only. Other enzymes have a protein part and a nonprotein part, both of which must be present before the enzyme can function. The protein part of such an enzyme is called the *apoenzyme*; the nonprotein part is called the *coenzyme*. The reaction is

$$\text{apoenzyme} + \text{coenzyme} \longrightarrow \text{enzyme}$$

Some vitamins act as coenzymes. Nicotinamide adenine dinucleotide, NAD, acts as a coenzyme in oxidation-reduction reactions in the mitochondria. Coenzyme A (CoA) functions in the metabolism of carbohydrates, lipids, and proteins.

a.  The nonprotein part of an enzyme is called the _____.

b.  The protein part of an enzyme is called the _____.

a.  **coenzyme**
b.  **apoenzyme**

One example of a coenzyme is _____.

**NAD or CoA or some vitamins**

# MODE OF ENZYME ACTIVITY

Enzymes are specific catalysts that contain an "active site," that section of the enzyme on which combination with the substrate takes place.

An older theory of enzyme activity, the *lock and key model*, assumes that the active site of an enzyme is rigid and that substrate molecules must fit into that site to react. Although it doesn't look like this illustration in reality, one small part of it might be represented like

Enzyme

If two compounds, A and B, approach the enzyme, they can attach themselves to it and then react together and go off as a compound, AB. The enzyme is left free to react again and is not used up.

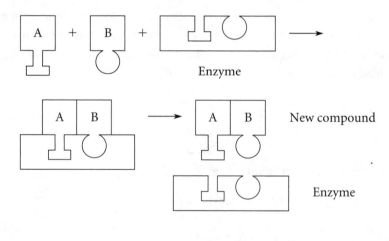

**specific; increase**

538

Enzymes are _____ (specific/nonspecific) catalysts that _____ the rate of chemical reaction in living systems.

Another version of the mode of enzyme activity is the *induced fit theory*. In this theory, the active site is flexible rather than rigid and it changes its conformation upon binding to the substrate, thus yielding an enzyme-substrate combination.

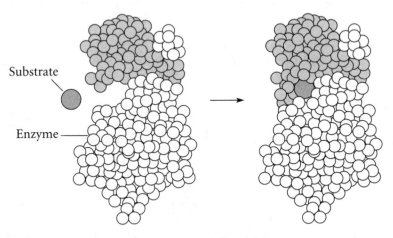

Substrate

Enzyme

Induced fit model

539

a.  In the lock and key model, the active site is _____ (flexible/rigid).

b.  In the induced fit model, the active site is _____ (flexible/rigid).

a. **rigid**
b. **flexible**

# INHIBITORS

540

If some substance other than the substrate fits into part of the active site, can the substrate then react with that enzyme? _____

**no**

a

no

### 541

A substance that competes for a position in the active site of an enzyme is called a *competitive inhibitor*. Which of the following substances could act as a competitive inhibitor for the active site shown in problem 542?

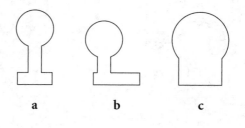

a          b          c

### 542

If substance **a** in problem 541 reacts with the active site of the enzyme shown here, could that enzyme then function normally? _____

Enzyme

543

An example of a competitive inhibitor is sulfanilamide, whose structure is similar to that of p-aminobenzoic acid, a compound that is essential for the growth of certain bacteria.

Sulfanilamide      p-Aminobenzoic acid

What effect should sulfanilamide, a competitive inhibitor, have on the growth of the bacteria that need p-aminobenzoic acid?

_____

it should inhibit the growth of those bacteria

544

Consider these diagrams:

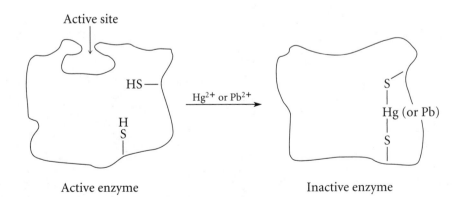

Active enzyme                              Inactive enzyme

Note that the shape of the active site has been changed. Can the enzyme then function normally? _____

no

$Hg^{2+}$ or $Pb^{2+}$

a. **competitive inhibitor**
b. **noncompetitive inhibitor**

A substance that changes the shape of the active site by reacting with the enzyme at some point other than the active site is called a *noncompetitive inhibitor.* What substance acts as a noncompetitive inhibitor in the reaction in problem 544? _____

546

a.  A substance that fits into the active site of an enzyme and prevents that enzyme from functioning normally is called a(n) _____.

b.  A substance that changes the shape of the active site of an enzyme by reacting with that enzyme at a position other than the active site is called a(n) _____.

# Methyl Groups  18

Methyl ($CH_3$—) groups are *not* functional groups, but they can impart specific biologic properties to some substances.

The following sections will refer to *methylamine* compounds only. For other types of methyl compounds, refer to an organic chemistry text.

Amines are compounds that contain an amine functional group ($NH_2$—).

Amines are named according to the number of carbon atoms present in the chain of carbons to which the amine is attached.

| | |
|---|---|
| $CH_3$—$NH_2$ | is called *methylamine*. |
| $CH_3$—$CH_2$—$NH_2$ | is called *ethylamine*. |
| $CH_3$—$CH_2$—$CH_2$—$NH_2$ | is called *propylamine*. |
| $CH_3$—$CH_2$—$CH_2$—$CH_2$—$NH_2$ | is called *butylamine*. |
| $CH_3$—$CH_2$—$CH_2$—$CH_2$—$CH_2$—$NH_2$ | is called *pentylamine*. |
| $CH_3$—$CH_2$—$CH_2$—$CH_2$—$CH_2$—$CH_2$—$NH_2$ | is called *hexylamine*. |

547

The prefix *methyl* indicates _____ (how many) carbon atoms.

1

548

2; 6

How many carbons are indicated by the prefix *ethyl*? _____
*Hexyl*? _____

butyl; pentyl

What prefix is used to indicate 4 carbon atoms in a chain?
_____ 5 carbons? _____

5

A chain of carbon atoms containing an amine group is numbered from one end to the other so that the amine group has the lowest possible number.

Thus, the following compound is numbered from right to left so that the $NH_2$ group is attached to carbon 1.

$$\overset{5}{C}H_3—\overset{4}{C}H_2—\overset{3}{C}H_2—\overset{2}{C}H_2—\overset{1}{C}H_2—NH_2$$

This compound is called *pentylamine*, where the number 1 is assumed and not written.

The prefix *pentyl* indicates how many carbon atoms in a chain?

_____

when no number is given, the number 1 is under-stood

Why is the number 1 not written?

_____

2; 5

**552**

The compound

$$\overset{5}{C}H_3—\overset{4}{C}H_2—\overset{3}{C}H_2—\overset{2}{C}H—\overset{1}{C}H_3$$
$$\underset{NH_2}{\shortmid}$$

is called 2-pentylamine (2-aminopentane), indicating that an amine group is present at carbon number _____ in a chain of how many carbon atoms? _____

### 553

Note the following aminopetane:

$$CH_3—CH_2—CH—CH_2—CH_3$$
$$|$$
$$NH_2$$

It is called _____.

3-pentylamine

### 554

In which direction are the carbon atoms numbered in problem 553?

_____

either direction

### 555

In which direction should the carbons be numbered? _____
Why? _____ What is the name of this compound?

_____

left to right; to give amine group lowest number; 2-pentylamine

### 556

The compound

$$CH_2—CH_2—CH_2—CH_2—CH_3$$
$$|$$
$$NH_2$$

should be called _____.

pentylamine with the number 1 being understood and the numbering from left to right a methyl group attached at carbon 2; a chain of 3 carbons; because no number is indicated, the amine group must be at carbon 1

## 557

If a methyl group is present in addition to an amino group, numbering of the carbons takes place continuously from one end of the molecule to the other so that the amine group has the lowest possible number. Consider the compound

If we number the carbon atoms from right to left, then the amine group is at carbon 1 (the lowest number) and a methyl group is located at carbon 2.

The name of this compound is 2-methylpropylamine, where 2-methyl indicates _____ and propyl indicates _____.

How do we know which carbon the amine group is attached to?

_____

one that contains 3 methyl groups

## 558

If two methyl groups are present in a molecule, then it is known as a dimethyl compound.

What is a trimethyl compound? _____.

2; methyl; 2 and 3; 4

## 559

If more than one methyl group is present in a compound, each is numbered separately, again using lowest possible numbers. In addition, a prefix (*di, tri, etc.*) is added to indicate the total number of methyl groups present.

Look at the following structure:

It is named 2,3-dimethylbutane, which indicates the presence of _____ (how many?) groups _____ (of which type?) at carbon _____ in a chain of _____ carbon atoms.

## 560

In which direction is the previous compound numbered?

either direction

_____

Consider the following two isomers:

$$CH_3—CH—CH_2—N—H \qquad CH_3—CH_2—CH_2—N—CH_3$$

with CH_3 below the CH, H below the N on compound a; and H below the N on compound b.

and

a                    b

Compound A is called 2-methylpropylamine. The *2-methyl* part indicates that there is a methyl group on carbon 2 of a 3-carbon chain (the propyl group) with an amine group at carbon 1 (number understood and not written).

Compound B is called *N-methylpropylamine*. The term *N-methyl* indicates that a methyl group is attached to a nitrogen. Again there is a 3-carbon group (propyl) with an amine group at carbon 1.

## 561

The compound

$$CH_3—CH—CH—CH_2—CH—CH_3$$

with CH_3, CH_3 below the second and third carbons, and CH_3 below the fifth carbon.

is called _____.

Note that if the numbering had taken place in the opposite direction, then the naming would have been 2,4,5-trimethylpentane with incorrect (higher) numbers.

2,3,5-trimethylhexane

**2,2,4-trimethylpentane**

**3,3-dimethylbutylamine;**
**N-N-dimethylbutylamine**

Caffeine

What is the name for the following compound?

$$CH_3-CH-CH_2-C-CH_3$$

with CH$_3$ above the CH and CH$_3$ groups above and below the C.

_____

Next consider the following two isomers:

c                                        d

Compound C is called _____.

Compound D is called _____.

Methyl groups are present in many biologically active compounds. Circle the methyl groups present in each of the following structures.

Caffeine

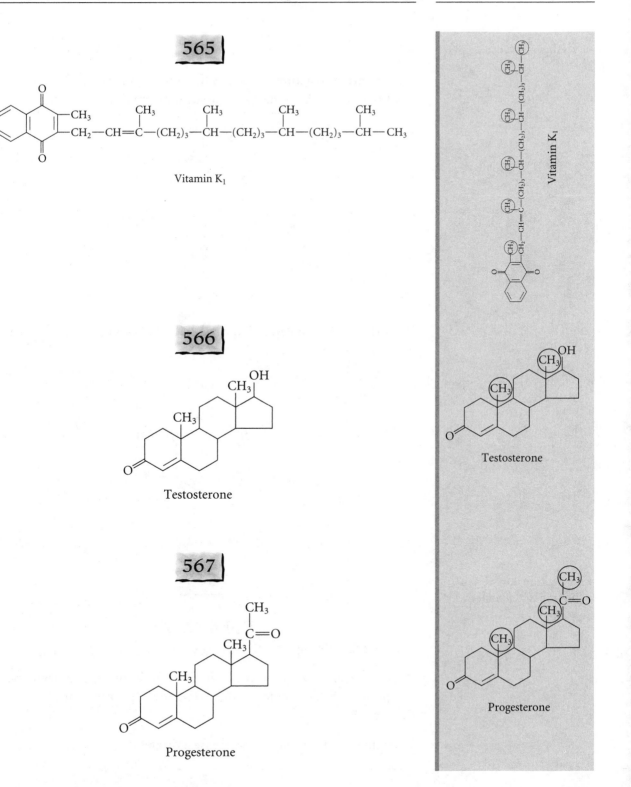

**565**

Vitamin K₁

**566**

Testosterone

**567**

Progesterone

**5-methylcytosine**

### 568

The following structure represents the compound *cytosine*, one of the DNA bases. Note how this structure is numbered.

Cytosine

If cytosine is methylated as indicated in the following structure, what would be its name and number? _____

**an enzyme involved in the transfer of a methyl group**

### 569

# ENZYMES

Enzymes usually have names ending in *ase*.

*Transferases* are enzymes involved in the transfer of a group from one compound (the donor) to another compound (the acceptor). See Chapter 17.

What does the term *methyltransferase* indicate?

_____

## 570

Consider the following abbreviated reaction:

Glycine          N-methylglycine

Note that a methyl group has been transferred to the glycine. The prefix *N-methyl* indicates that the methyl group is attached where on the glycine molecule? _____

## 571

The common name for the enzyme involved in this reaction is *glycine N-methyltransferase*.

The ending _____ indicates that it is an enzyme.

## 572

The word *transferase* indicates that this enzyme is involved in what type of reaction? _____

## 573

Methyltransferase indicates the transfer of which group? _____

---

to the nitrogen atom

ase

a transfer reaction

a methyl group

that the methyl group
has been transferred to
the nitrogen of the
glycine

**574**

The N indicates_____.

acceptor

**575**

The word *glycine* at the beginning of an enzyme's name indicates that
the substance is either the donor or the acceptor of the methyl group. A
careful examination of the sturctures of glycine and N-methylglycine
indicates that the glycine must be the _____ (donor/acceptor).

a methyl group

**576**

Consider the enzyme with the common name
*nicotinate N-methyltransferase*. The reaction is

    nicotinate + methionine ⟶ homocysteine + N-methylnicotinate

What group is being transferred? _____

nicotinate

**577**

To which compound is the methyl group being transferred?

_____

acceptor

**578**

Nicotinate is the _____ (donor/acceptor) substance.

a methyl group

**579**

If an enzyme has the common name ending in *S-methyltransferase*, then
what group is being transferred? _____

## 580

Where is that methyl group being transferred to? _____

to a sulfur atom

## 581

What does the *O* indicate in the following common name of an enzyme?

*acetylserotonin O-methyltransferase*

_____

transfer of a methyl group to an oxygen atom

## 582

# DNA METHYLTRANSFERASE

DNA methylation occurs when a methyl group or groups is/are attached to a DNA base after that DNA has been synthesized.

DNA methylation in general inactivates a gene produced by a highly methylated DNA base. Such a gene is said to be "turned off."

In DNA methylation, DNA is a(n) _____ (donor/acceptor).

acceptor

activation of a gene or turning a gene "on"

## 583

# DEMETHYLATION

Demethylation occurs in the presence of enzymes called *demethylases*.

If methylation of *DNA* can cause the inactivation of a gene, then what effect might demethylation produce? _____

# REVIEW QUESTIONS

What is a transferase? _____

an enzyme involved in the transfer of a group during a reaction

What is a methyltransferase?

_____

an enzyme involved in the transfer of a methyl group

**586**

What is an N-methyltransferase?

_____

an enzyme involved in the transfer of a methyl group to an N atom in an acceptor

**587**

What is an S-methyltransferase?

_____

an enzyme involved in the transfer of a methyl group to an S atom in an acceptor

**588**

What is an O-methyltransferase?

_____

an enzyme involved in the transfer of a methyl group to an O atom in an acceptor

What is a DNA methyltransferase?

_____

an enzyme involved in the transfer of a methyl group (or methyl groups) to DNA

In DNA methyltransferase, what is the function of the DNA?

_____

to act as an acceptor of methyl(s)

591

What happens to a gene produced by a highly methylated DNA base?

_____

it is turned "off"

592

What are the names of the following compounds?

a. CH₃—CH—CH₂—CH₂—NH₂
        |
        CH₃

b.

c. CH₃—CH₂—CH₂—N—CH₃
                |
                CH₃

a. 3-methylbutylamine
b. 2,2-dimethylpropane
c. N,N-dimethylpropy-
   lamine

# Biologic Oxidation 19

Oxidation may be defined as

🌿 a loss of an electron (or electrons)

🌿 a gain of oxygen

🌿 a loss of hydrogen

🌿 an increase in oxidation number

The compound lactic acid may be oxidized to pyruvic acid according to the following equation.

Lactic acid                Pyruvic acid

$$\boxed{593}$$

This reaction is an oxidation reaction because it involves:

__ the gain of oxygen by the lactic acid molecule
__ the loss of hydrogen by the lactic acid molecule

the loss of hydrogen by
the lactic acid molecule

**a. hydroxyl and carboxyl**
**b. carbonyl and carboxyl**

**594**

a.   Lactic acid contains what functional groups? _____

b.   Pyruvic acid contains what functional groups? _____

**gain of oxygen**

**595**

Another example of oxidation is the reaction of acetaldehyde to form acetic acid:

$$
\underset{\text{Acetaldehyde}}{
\begin{array}{c}
\text{H}\quad\text{H}\\
|\qquad|\\
\text{H}-\text{C}-\text{C}=\text{O}\\
|\\
\text{H}
\end{array}}
\xrightarrow{\text{[O]}}
\underset{\text{Acetic acid}}{
\begin{array}{c}
\text{H}\quad\text{OH}\\
|\qquad|\\
\text{H}-\text{C}-\text{C}=\text{O}\\
|\\
\text{H}
\end{array}}
$$

This reaction is oxidation because it involves _____.

**there is a loss of**
**an electron**

**596**

The $Fe^{2+}$ ion may be oxidized to the $Fe^{3+}$ ion as follows:

$$Fe^{2+} - e^- \longrightarrow Fe^{3+}$$

Why is it oxidation? _____

Reduction is the opposite of oxidation. Reduction is

-   a gain of an electron (or electrons)

-   a gain of hydrogen

-   a loss of oxygen

-   a decrease in oxidation number

Oxidation and reduction always occur together. One cannot occur without the other.

## 597

Consider the following reaction:

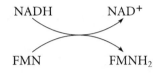

What was oxidized? _____ Why? _____

What was reduced? _____ Why? _____

Note that both oxidation and reduction are involved in this reaction.

## 598

Consider the following reaction:

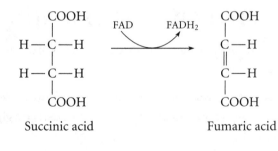

Succinic acid                    Fumaric acid

a.  Is succinic acid oxidized or reduced? _____
    Why? _____

b.  Is FAD oxidized or reduced? _____ Why? _____

NADH; it lost hydrogen;
FMN; it gained hydrogen

a.  oxidized; there is a
    loss of hydrogen
b.  reduced; there is a
    gain of hydrogen

FAD was reduced; it
gained hydrogens
succinate was oxidized; it
lost hydrogens

Oxidation-reduction reactions in living organisms take place primarily between ions in solution.

The molecular reaction in problem 598 may be written in ionic form as

Succinate                              Fumarate

where the word *ion* is understood and not written.

What was reduced? _____     Why? _____

What was oxidized? _____     Why? _____

# CELLULAR RESPIRATION

Cellular respiration enables cells to produce energy from food through oxidation-reduction reactions.

Look at Figure A in the Appendix and refer to it for problems 600–623. Note that cellular respiration involves three distinct stages: glycolysis, the citric acid cycle, and the electron transport chain.

cytoplasmic fluid;
mitochondrion;
mitochondrion

In which part of the cell does glycolysis take place? _____

In which part of the cell does the citric acid cycle occur?_____

In which part of the cell does the electron transport chain function?

_____

## 601

Glycolysis refers to the conversion of glucose to _____ and takes place in which part of the cell? _____

pyruvate, cytoplasmic fluid

## 602

Pyruvate from the glycolysis sequence is changed to acetyl CoA and enters the _____ cycle, which takes place in the _____ of the cell.

citric acid; mitochondrion

## 603

What substance transfers electrons from the glycolysis sequence to the electron transport chain? _____

NADH

## 604

Where does the NADH from the glycolysis sequence go? _____

to the electron transport chain

## 605

Where else is NADH produced? _____

in the citric acid cycle and in conversion of pyruvate to acetyl CoA

## 606

During glycolysis, glucose is converted to _____.

pyruvate

cytoplasmic fluid

### 607

Glycolysis takes place in the _____ of the cell.

glucose

### 608

In glycolysis, _____ is converted to pyruvate.

electron transport chain;
NADH

### 609

Electrons are transferred from the glycolysis sequence to the _____
via what substance? _____

mitochondrion

### 610

The electron transport chain functions in which part of the cell?
_____

mitochondrion

### 611

The citric acid cycle functions in which part of the cell? _____

3

### 612

Glycolysis produces pyruvate, NADH, and one other substance,
ATP, the cell's main energy compound.

The abbreviation ATP refers to adenosine triphosphate. This name
indicates that the compound adenosine is bonded to _____
(how many?) phosphate groups.

## 613

The abbreviation ADP refers to adenosine diphosphate, indicating that the adenosine molecule is bonded to _____ (how many?) phosphate groups.

## 614

The abbreviation AMP refers to adenosine monophosphate and indicates that the _____ molecule is bonded to _____ (how many?) _____ group(s).

## 615

Pyruvate from the glycolysis sequence enters the citric acid cycle after being changed into _____.

## 616

Electrons are transported from the citric acid cycle via _____ and _____.

## 617

Electrons are transported from the citric acid cycle to the _____.

## 618

The electron transport chain functions in which part of the cell? _____

2

adenosine; 1; phosphate

acetyl CoA

NADH; FADH$_2$

electron transport chain

mitochondrion

glycolysis; conversion of
pyruvate to acetyl CoA;
citric acid cycle

## 619

NADH is produced in three different sequences during cellular respiration. They are _____, _____, and _____.

citric acid cycle

## 620

$FADH_2$ is produced during which sequence? _____

ATP

## 621

The citric acid cycle produces the energy compound _____.

ATP

## 622

The electron transport chain produces the energy compound _____.

glycolysis; citric acid
cycle; electron transport
chain

## 623

During cellular respiration, ATP is produced in the following sequence(s): _____, _____, and _____.

# GLYCOLYSIS

Refer to Figure B in the Appendix for problems 624–643.

glucose; 6

## 624

Glycolysis begins with _____ (which compound?), which contains _____ (how many?) carbon atoms.

## 625

The end product of glycolysis is _____, which contains
_____ (how many?) carbon atoms.

pyruvate; 3

## 626

For each molecule of glucose that goes through the glycolysis sequence,
how many molecules of pyruvate are produced? _____

2

## 627

In which part of the cell does glycolysis take place? _____

cytoplasmic fluid

## 628

Note that in the first half of the glycolysis sequence, ATP is used up
and is converted to _____.

ADP

## 629

How many ATPs are required for the first three steps in the glycolysis
sequence? _____

The ATP thus used is converted into _____.

2; ADP

## 630

In the second half of the glycolysis sequence, energy is produced
and ADP is converted to _____.

ATP

4

## 631

In the second half of the glycolysis sequence, _____ (how many?) ADPs are converted to ATPs.

2

## 632

In the first half of the glycolysis sequence, _____ (how many?) ATPs are converted to ADPs.

2

## 633

In the entire glycolysis sequence, the net number of ATPs produced is _____.

2

## 634

In addition to ATP, another energy compound, NADH, is produced during glycolysis. How many NADHs are produced from one molecule of glucose during the glycolysis sequence? _____

NAD$^+$

## 635

NADH is produced from which compound? _____

pyruvate

## 636

The overall glycolysis sequence may be written as

glucose  _____

### 637

The overall glycolysis sequence may be further written as

$$\text{glucose} \longrightarrow \underline{\hspace{2cm}} \text{ (how many?) pyruvate}$$

### 638

A further addition to this equation is

```
        ADP          _____
           \
glucose  ————\————→ 2 pyruvate
```

### 639

This equation may then be modified to

```
      2 ADP      ____ (how many?) ATP
          \
glucose ———\———→ 2 pyruvate
```

### 640

This equation may be further modified to

```
      2 ADP          2 ATP
          \           /
glucose ———\———————→ 2 pyruvate
          /           \
      2 NAD⁺           2 _____
```

2

ATP

2

NADH

a. **2 ADP**
b. **2 ATP**
c. **2 pyruvate**
d. **2 NAD$^+$**
e. **2 NADH**

**no**

**anaerobic**

**pyruvate**

The complete glycolysis sequence may thus be written as

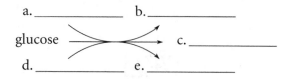

a. _____    b. _____

glucose

c. _____

d. _____    e. _____

A biochemical reaction that requires oxygen is called *aerobic*; one that does not require oxygen is called *anaerobic*.

Do any of the reactions in the glycolysis sequence require oxygen?

_____

643

Is glycolysis an aerobic or an anaerobic sequence? _____

## CONVERSION OF PYRUVATE TO ACETYL COENZYME A

Refer to Figure C in the Appendix for problems 644–657.

The end product of the glycolysis sequence is _____.

## 645

Pyruvate is produced in which part of the cell? _____

cytoplasmic fluid

## 646

Pyruvate from the glycolysis sequence diffuses from the cytoplasmic fluid into which part of the cell? _____

mitochondrion

## 647

Does pyruvate enter the citric acid cycle directly? _____

no

## 648

Pyruvate is converted into which compound in the mitochondrion?

_____

acetyl CoA

## 649

Does acetyl CoA (acetyl coenzyme A) enter the citric acid cycle?

_____

yes

## 650

Each molecule of glucose that enters the glycolysis sequence produces _____ (how many?) molecule(s) of pyruvate.

2

1

2; 2

6; 6; 2

2; pyruvate

## 651

Each pyruvate that diffuses from the cytoplasmic fluid into the mitochondrion produces _____ (how many?) molecule(s) of acetyl CoA.

## 652

Therefore, during glycolysis, one molecule of glucose produces _____ pyruvate(s), which in turn are used to produce _____ molecule(s) of acetyl CoA.

## 653

One molecule of glucose contains _____ (how many?) carbon atoms.

Two molecules of pyruvic acid contain _____ (how many?) carbon atoms.

The acetyl groups of two molecules of acetyl CoA contain four carbon atoms.

How many carbon atoms are missing during this overall set of reactions? _____

## 654

As is shown in Figure C in the Appendix, in the cytoplasmic fluid, one molecule of carbon dioxide is produced for each molecule of pyruvate converted into acetyl CoA. Thus, one molecule of glucose enters the glycolysis sequence and is converted to _____ (how many?) molecule(s) of _____.

## 655

In the mitochondrion, two molecules of pyruvate are converted into _____ (how many?) molecule(s) of _____, which then enter the citric acid cycle.

2; acetyl CoA

## 656

The missing carbon atoms show up as _____.

carbon dioxide or $CO_2$

## 657

For each molecule of acetyl CoA produced from pyruvic acid, _____ (how many?) molecule(s) of NADH are produced. How many molecules of $CO_2$ are produced? _____

1; 1

# THE CITRIC ACID CYCLE

Refer to Figure D in the Appendix for problems 658–668.

## 658

The two carbon atoms of the acetyl part of acetyl CoA participate in the citric acid cycle while the CoA molecule is recycled.

The two carbon atoms from the acetyl group are eliminated from the citric acid cycle as two molecules of what compound? _____

$CO_2$

## 659

As one molecule of acetyl CoA goes through the citric acid cycle:

How many molecules of ATP are produced? _____

How many molecules of NADH are produced? _____

How many molecules of $FADH_2$ are produced? _____

1; 3; 1

1; 3; 1

### 660

For every molecule of acetyl CoA that enters the citric acid cycle, _____ molecule(s) of ATP, _____ molecule(s) of NADH, and _____ molecule(s) of $FADH_2$ are produced.

ADP

### 661

In the citric acid cycle, ATP is produced from what compound?
_____

NAD$^+$

### 662

In the citric acid cycle, NADH is produced from what compound?
_____

FAD

### 663

In the citric acid cycle, $FADH_2$ is produced from what compound?
_____

mitochondrion

### 664

In which part of the cell does the citric acid cycle take place?
_____

$CO_2$; ATP; NADH; $FADH_2$

### 665

The four products eliminated from the citric acid cycle are
_____, _____, _____, and _____.

Is oxygen required in any of the steps of the citric acid cycle?

_____

no

---

Is the citric acid cycle an aerobic or anaerobic sequence? _____

anaerobic

---

Where do the NADH and the FADH$_2$ produced by the citric acid cycle go? _____

to the electron transport chain

## ELECTRON TRANSPORT CHAIN

Refer to Figure E in the Appendix for problems 669–680.

---

The reactions in the electron transport chain take place in which part of the cell? _____

mitochondrion

---

Is oxygen required in any of the steps of the electron transport chain?

_____

yes

---

Is the electron transport chain an aerobic or anerobic sequence?

_____

aerobic

NADH, FADH$_2$

Which compounds enter the electron transport chain? _____

citric acid cycle

The FADH$_2$ entering the electron transport chain comes from which sequence? _____

glycolysis, conversion of pyruvate to acetyl CoA, and the citric acid cycle

The NADH entering the electron transport chain comes from three different sequences. They are _____

_____ .

ATP

What is the primary energy compound produced by the electron transport chain? _____

yes

Is ATP produced during glycolysis? _____

no

677

Is ATP produced during the conversion of pyruvate to acetyl CoA?

_____

## 678

Is ATP produced in the citric acid cycle? _____

yes

## 679

Is ATP produced in the electron transport chain? _____

yes

## 680

ATP is produced from which compound? _____

ADP

# Photosynthesis 20

## OVERALL REACTION

### 681

The overall reaction of photosynthesis is

$$6CO_2 + 6H_2O \xrightarrow[\text{chlorophyll}]{\text{light}} C_6H_{12}O_6 + 6O_2$$

During photosynthesis,

a. the raw materials are _____

b. the end products are _____

c. the source of energy is _____

d. the source of $CO_2$ is _____

e. the source of $H_2O$ is _____

f. what happens to the $O_2$? _____

a. $CO_2$ and $H_2O$
b. $C_6H_{12}O_6$ and $O_2$
c. light
d. air
e. the ground
f. released into air

**animal; plant**

682

Look at the following structures for heme (which is a part of the hemoglobin molecule) and chlorophyll *a*.

Heme

Chlorophyll *a*

Note that these two substances have similar structures. Heme (hemoglobin), however, is necessary for most _____ (animal/plant) life, whereas chlorophyll *a* is necessary for most _____ (animal/plant) life.

## 683

What metal is at the center of the heme molecule? _____ The
chlorophyll *a* molecule? _____

## 684

Consider the following two examples of the overall photosynthesis reaction. Radioactive oxygen, $^{18}O$, is introduced in the form of radioactive
$C^{18}O_2$ in the first reaction and as radioactive water, $H_2^{18}O$, in the second
reaction.

$$6C^{18}O_2 + 6H_2O \xrightarrow[\text{chlorophyll}]{\text{light}} C_6H_{12}^{18}O_6 + 6O_2$$

$$6CO_2 + 6H_2^{18}O \xrightarrow[\text{chlorophyll}]{\text{light}} C_6H_{12}O_6 + 6^{18}O_2$$

What is the source of the $O_2$ produced during photosynthesis, $CO_2$ or
$H_2O$?

## 685

Photosynthesis may be subdivided into two main steps as indicated in
the following simplified chart.

a. light
b. $H_2O$
c. ATP, NADPH; $O_2$
d. $H_2O$

a. $CO_2$
b. carbohydrate or
$C_6H_{12}O_6$
c. ADP; NADP$^+$
d. no

For the first step, *light reactions,*

a.   what is the energy source? _____

b.   what reactant is necessary? _____

c.   what compounds are produced? _____

d.   where does the $O_2$ come from? _____

# 686

For the second step, the *Calvin cycle,*

a.   what is the principal reactant? _____

b.   what is the principal product? _____

c.   what energy compounds are returned to the light reactions step?

_____

d.   Is light required for the Calvin cycle? _____

# LIGHT REACTIONS

The following is a simplifed diagram of light reactions.

When light strikes the chlorophyll molecules, it excites an electron, which is transferred from chlorophyll molecule to chlorophyll molecule until it reaches the reaction center. In photosystem I, the reaction center is associated with a specific protein (pigment) known as P700 because it is most effective at absorbing light with a wavelength of 700 nm. Photosystem II is associated with a similar protein P680 that is sensitive to absorbing light with a _____.

688

Which photosystem has a noncyclic electron flow? _____

**wavelength of 680 nm**

**II**

P700; P680

I; II

1

yes

What numbered form of chlorophyll *a* is associated with photosystem I? _____ With photosystem II? _____

Which photosystem is cyclic? _____ Noncyclic? _____

The following is an abbreviated diagram of photosystem I.

How many electrons are involved in this system? _____

In photosystem I, an electron is emitted by P700 and travels to a primary acceptor. After a series of reactions, does the electron return to P700?

_____

That is why this system is called cyclic.

## 693

Is $H_2O$ split to provide electrons in photosystem I? _____

no

## 694

Is $O_2$ produced during the reactions of photosystem I? _____

no

## 695

What product is the result of the reactions in photosystem I?

_____

ATP

## 696

In photosystem I, ATP is produced from _____.

ADP

## 697

When ADP is converted to ATP, what group must be added?

_____

P (or phosphate)

## 698

The production of ATP during the cyclic electron flow is called *cyclic photophosphorylation*.

During cyclic photophosphorylation, _____ is converted to

_____.

ADP; ATP

no

no

P700; light

yes; no;
yes; yes

---

## 699

Is NADPH produced in this sequence? _____

## 700

Is $O_2$ produced in this sequence? _____

## 701

The energy for the electron involved in photosynthesis I is produced from the chlorophyll *a* that is labeled _____ and in turn is produced by the action of _____ on chlorophyll.

The following is an abbreviated diagram of photosystem II.

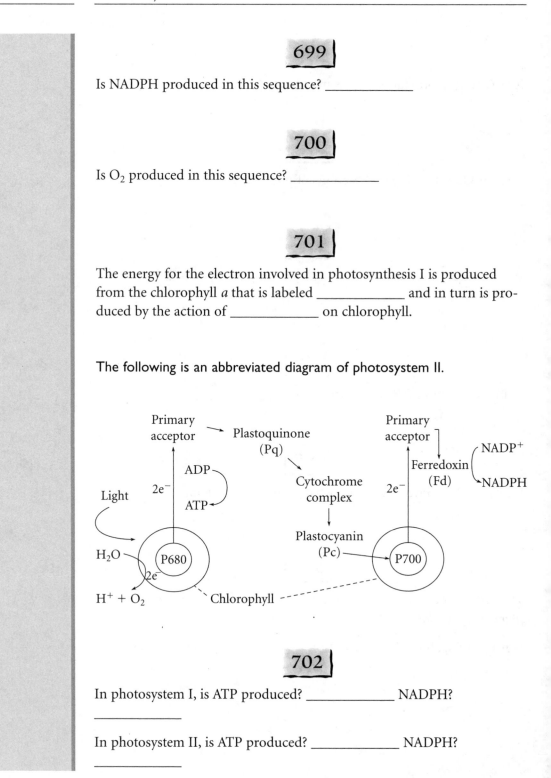

## 702

In photosystem I, is ATP produced? _____ NADPH? _____

In photosystem II, is ATP produced? _____ NADPH? _____

## 703

Where do the electrons used by photosystem II come from?

_____

water

## 704

When the $H_2O$ gives off electrons, what substances are produced?

_____

$H^+, O_2$

## 705

Is the same type of chlorophyll *a* needed for both photosystems?
(yes/no) _____ Why?

no, system I needs P700
and system II needs P680

## 706

Can ATP be produced in both photosystems? _____

yes

## 707

Is photosystem II cyclic? _____

no

## 708

Is water splitting required for photosystem I? _____ For photosystem II? _____

no; yes

## 709

Is $O_2$ produced in photosystem I? _____ In photosystem II?

_____

no; yes

no; yes

NADP$^+$

$H_2O$

light

## 710

Is NADPH produced in photosystem I? _____ In photosystem II? _____

## 711

From what substance is NADPH produced? _____

## 712

From what substance is $O_2$ produced? _____

The light reactions system may be abbreviated as shown in the following diagram.

## 713

What is the original energy source for this series of reactions?

_____

## 714

When light strikes photosystem I and photosystem II, it causes electrons to flow.

Which of these systems is cyclic? _____

Which is noncyclic? _____

I; II

## 715

In which photosystem is water splitting involved? _____

II

## 716

What is the source of oxygen during the light reactions? _____

water

## 717

By which photosystem is oxygen produced? _____

II

## 718

What high-energy compound(s) is(are) produced in photosystem I? _____

ATP

## 719

What high-energy compound(s) is(are) produced in photosystem II? _____

ATP, NADPH

**ADP**

**NADP⁺**

From what substances is ATP produced? _____

From what substance(s) is NADPH produced? _____

## CALVIN CYCLE

The following is an abbreviated diagram of the Calvin cycle.

The end product of the Calvin cycle is glyceraldehyde-3-P. It contains how many carbon atoms? _____

## 723

$CO_2$ contains how many carbon atoms? _____

1

## 724

How many $CO_2$ molecules must go through the Calvin cycle to produce one molecule of glyceraldehyde-3-P? _____

3

## 725

Is $CO_2$ a part of the light reactions sequence or the Calvin cycle? _____

Calvin cycle

## 726

What high-energy compounds enter the Calvin cycle? _____

ATP, NADPH

## 727

Where do these energy compounds come from? _____

light reactions

The following is a simplified illustration of the Calvin cycle. Numbers in brackets [ ] indicate total number of carbon atoms. Enzymes for each step are omitted for simplicity.

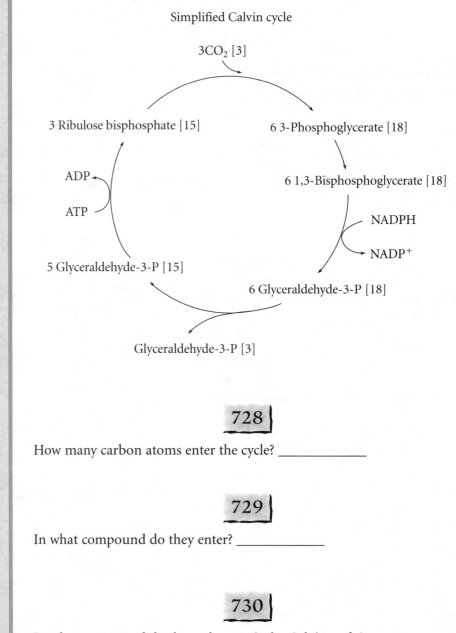

Simplified Calvin cycle

$3CO_2$ [3]

3 Ribulose bisphosphate [15]

6 3-Phosphoglycerate [18]

ADP

ATP

6 1,3-Bisphosphoglycerate [18]

NADPH

NADP$^+$

5 Glyceraldehyde-3-P [15]

6 Glyceraldehyde-3-P [18]

Glyceraldehyde-3-P [3]

**3**

**$CO_2$**

**glyceraldehyde-3-P**

> 728

How many carbon atoms enter the cycle? _____

> 729

In what compound do they enter? _____

> 730

In what compound do the carbons exit the Calvin cycle? _____

### 731

Glyceraldehyde-3-P contains how many carbons? _____

3

### 732

Does all the glyceraldehyde-3-P produced in the Calvin cycle leave the cycle? _____

no

### 733

What fraction of the glyceraldehyde-3-P leaves that cycle?

_____

⅙

### 734

Does the number of carbon atoms entering, reacting in, and leaving the cycle balance? _____

yes

### 735

What energy compounds enter the Calvin cycle? _____

ATP, NADPH

### 736

What energy compounds are produced in the Calvin cycle?

_____

ADP, NADP$^+$

**ATP; NADPH**

### 737

These energy compounds return to the light reactions system. There, ADP is regenerated to _____ and NADP$^+$ is regenerated to _____.

**3**

### 738

Glyceraldehyde-3-P contains how many carbon atoms? _____

**6**

### 739

Glucose, $C_6H_{12}O_6$, contains how many carbon atoms? _____

**2**

### 740

How many glyceraldehyde molecules must be combined to produce one glucose (carbohydrate) molecule? _____

The following figure gives an overall summary of photosynthesis.

Summary of photosynthesis

$\boxed{741}$

What is the substance A that enters the light reactions system?

_____

$\boxed{742}$

What is substance B? _____

$\boxed{743}$

What compounds are represented by E? _____

$\boxed{744}$

What compounds are represented by F? _____

$\boxed{745}$

What is compound C? _____

$\boxed{746}$

What is compound D? _____

$H_2O$

$O_2$

**ATP, NADPH**

**ADP, NADP$^+$**

$CO_2$

**glyceraldehyde-3-P**

# C$_4$ PLANTS

C$_4$ plants are those that have an alternate method of fixing $CO_2$ for the Calvin cycle as indicated in the following abbreviated diagram. (Numbers in brackets [   ] refer to the number of carbon atoms in each molecule.)

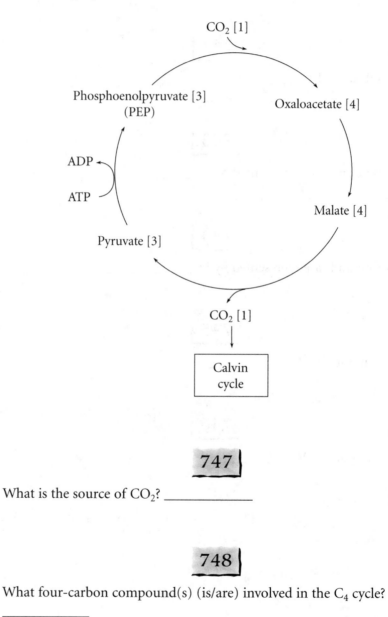

air

748

748

---

747

What is the source of $CO_2$? _____

malate, oxaloacetate

748

What four-carbon compound(s) (is/are) involved in the C$_4$ cycle?

_____

**749**

What high-energy compound is needed for the $C_4$ cycle? _____

ATP

**750**

When $CO_2$ enters the Calvin cycle, what high-energy compound(s) (is/are) required? _____

ATP, NADPH

**751**

Where are these high-energy compounds produced? _____

**in light reactions sequence**

# Oxygen–Carbon Dioxide Transport in Blood  21

## FLOW OF GASES

Gases diffuse from an area of higher pressure to one of lower pressure.

In the following diagram, $P_{O_2}$ indicates the partial pressure of oxygen (that part of the total pressure exerted by the oxygen) in millimeters of mercury.

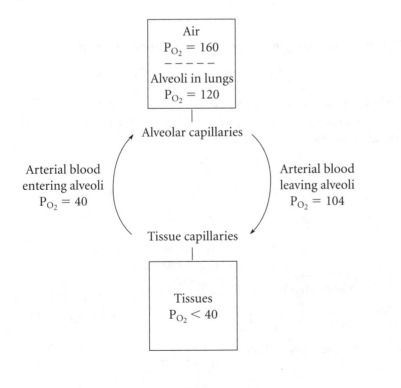

Air
$P_{O_2} = 160$
– – – – –
Alveoli in lungs
$P_{O_2} = 120$

Alveolar capillaries

Arterial blood
entering alveoli
$P_{O_2} = 40$

Arterial blood
leaving alveoli
$P_{O_2} = 104$

Tissue capillaries

Tissues
$P_{O_2} < 40$

Which way does oxygen diffuse in the lungs? _____

**into the capillaries in the alveoli**

---

From the alveoli, oxygen diffuses into the blood and then to the tissues. Why does it move in that direction? _____

**gases diffuse from higher to lower partial pressure**

---

**754**

When the arterial blood flows back to the alveoli, does it pick up or give off oxygen? _____

**picks up**

---

Tissues require oxygen for metabolic reactions such as

$$C_6H_{12}O_6 + 6O_2 \longrightarrow 6CO_2 + 6H_2O$$

Carbon dioxide is a gaseous waste product of oxidative metabolism. Its partial pressures in various parts of the blood system are shown in the following figure.

From the tissues, which way does carbon dioxide flow? _____

_____

by diffusion to venous blood to arterial blood to alveoli to lungs to air

Why does carbon dioxide flow in that direction? _____

_____

gases flow from an area of higher partial pressure to one of lower partial pressure

# TRANSPORT OF OXYGEN

### 757

When oxygen enters the red blood cells, it binds with hemoglobin, here represented by the formula HHb, to form a compound known as oxy-hemoglobin.

$$HHb + O_2 \longrightarrow HHbO_2$$

Hemoglobin              Oxyhemoglobin

Oxyhemoglobin in turn is ionized into the oxyhemoglobin ion and a hydrogen ion.

$$HHbO_2 \longrightarrow H^+ + HbO_2^-$$

Oxygen is carried to the tissues in the form of the oxyhemoglobin ion, or _____.

$HbO_2^-$

### 758

In the tissues, the oxyhemoglobin ion binds with a hydrogen ion to form oxyhemoglobin; the formula is

$$HbO_2^- + H^+ \longrightarrow \underline{\hspace{2cm}}$$

$HHbO_2$

$$HHbO_2 \longrightarrow HHb + O_2$$

## 759

The oxyhemoglobin releases its oxygen to form hemoglobin. This reaction may be written as _____.
The oxygen thus liberated is used by the tissues for various metabolic reactions. The hemoglobin once again is available to pick up oxygen in the alveoli.

**hemoglobin**

## 760

What substance in the blood carries oxygen to the tissues? _____

**HHb**

## 761

What is the symbol for hemoglobin? _____

**oxyhemoglobin**

## 762

When hemoglobin picks up oxygen, _____ is formed.

**HHbO$_2$**

## 763

What is the formula of oxyhemoglobin? _____

$$HHb + O_2 \longrightarrow HHbO_2$$

## 764

Write the reaction for hemoglobin and oxygen. _____
_____

## 765

Write the formula for the ionization of oxyhemoglobin. _____

$$HHbO_2 \rightarrow H^+ + HbO_2^-$$

## 766

In what form is oxygen carried to the tissues? _____

$HbO_2^-$ or oxyhemoglobin ion

## 767

Write the reaction of the oxyhemoglobin ion with a hydrogen ion in the tissues. _____

$$HbO_2^- + H^+ \rightarrow HHbO_2$$

## 768

Write the reaction indicating how oxyhemoglobin gives off its oxygen in the tissues. _____

$$HHbO_2 \rightarrow HHb + O_2$$

## 769

What happens to the hemoglobin after it gives off its oxygen?

returns to alveoli to pick up more $O_2$

**red blood cells**

770

What part of the blood contains hemoglobin? _____

The overall steps in this process may be diagrammed as follows.

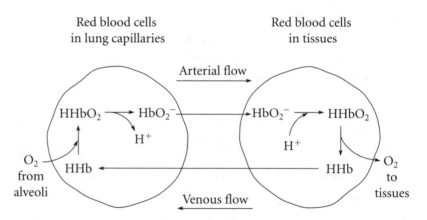

# TRANSPORT OF CARBON DIOXIDE

**gases diffuse from an area of higher to lower partial pressure**

771

Carbon dioxide is produced in the tissues and diffuses into the red blood cells.

Why does it diffuse in that direction? _____
_____

## 772

In the red blood cells, carbon dioxide reacts with water to produce carbonic acid:

$$CO_2 + H_2O \longrightarrow H_2CO_3$$
Carbonic acid

Carbonic acid in turn ionizes, producing a hydrogen ion and a bicarbonate ion.

$$H_2CO_3 \longrightarrow H^+ + HCO_3^-$$
Bicarbonate
ion

Red blood cells are able to hold only a small amount of the bicarbonate ions produced from carbon dioxide, so the excess diffuses into the blood plasma.

The bicarbonate ion has what charge? _____

## 773

To replace the bicarbonate ions, chloride ions ($Cl^-$) flow from the plasma into the red blood cells.

Chloride ions have what charge? _____

## 774

The process of shifting bicarbonate and chloride ions in opposite directions across a red blood cell membrane is called the *chloride shift*. It takes place in the red blood cells in the lung and tissue capillaries.

In the chloride shift, _____ ions flow from the red blood cells to the _____.

−1

−1

$HCO_3^-$; plasma

Cl$^-$; plasma; red blood cells

### 775

As bicarbonate ions flow out of the red blood cells, _____ ions flow from the _____ into the _____.

H$_2$CO$_3$  →  H$_2$O + CO$_2$

### 776

In the lungs, the bicarbonate ion combines with a hydrogen ion to re-form carbonic acid.

$$HCO_3^- + H^+ \longrightarrow H_2CO_3$$

Carbonic acid in turn decomposes into water and carbon dioxide, which is then exhaled. The equation for this reaction is _____.

most of it is exhaled

### 777

What happens to the carbon dioxide in the lungs? _____

chloride; bicarbonate

### 778

As the bicarbonate ion is used up in the preceding reactions, it is replaced by bicarbonate ions from the plasma. At the same time, chloride ions from the red blood cells flow back into the plasma. This process is termed a *reverse chloride shift*.

In a reverse chloride shift, _____ ions flow from the red blood cells into the plasma and at the same time _____ ions flow from the plasma into the red blood cells.

These reactions may be diagrammed as follows.

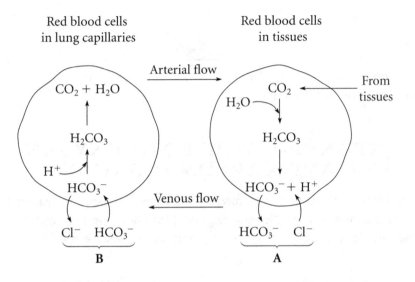

What does part **A** represent? _____

What does part **B** represent? _____

In addition to the red blood cells transporting carbon dioxide as bicarbonate ions, some of the carbon dioxide reacts with hemoglobin to form carbaminohemoglobin:

$$CO_2 + HHb \longrightarrow HHbCO_2$$

In the lungs, carbaminohemoglobin decomposes into hemoglobin and carbon dioxide.

What happens to the carbon dioxide? _____

What happens to the hemoglobin? _____

**hemoglobin**

Oxygen is carried to the tissues by what compound? _____

**bicarbonate; carbamino-
hemoglobin**

## 783

Carbon dioxide is carried from the tissues in the form of _____ ions and also in the compound _____.

# TRANSPORT OF BOTH OXYGEN AND CARBON DIOXIDE

The interrelationship of the oxygen and carbon dioxide transport systems is shown in the following diagram. Note how the hydrogen ions in the red blood cells in both the lungs and the tissues are recycled.

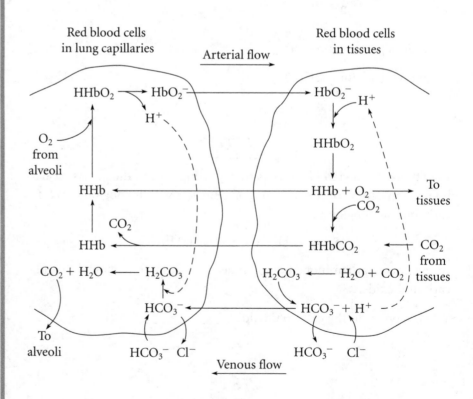

# Review

## 784

What elements do the following symbols represent?

a. H _____     b. O _____     c. S _____

d. Na _____     e. K _____     f. Fe _____

## 785

Write the symbol for each of the following elements:

a. chlorine _____     b. zinc _____     c. calcium _____

d. carbon _____     e. bromine _____     f. magnesium _____

## 786

When an atom gains an electron, it forms an ion with what charge?

_____

## 787

The chlorine atom gains 1 electron to form a chloride ion. The symbol for the chloride ion is _____.

**ionic**

$$\cdot \; \overset{\displaystyle \cdot}{\underset{\displaystyle \cdot}{C}} \; \cdot$$

$$: \overset{..}{O} :: C :: \overset{..}{O} :$$

**hydrogen (or H$^+$)**

**hydrogen (or H$^+$)**

**any ions except H$^+$
and OH$^-$**

### 788

When two ions are held together by the attraction of their opposite charges, what type of bond is between them? _____

### 789

Draw the Lewis dot structure for carbon, atomic number 4.

### 790

Draw the Lewis dot structure for $CO_2$.

### 791

An acid is a substance that yields _____ ions in solution.

### 792

A base is a substance that accepts _____ ions in solution.

### 793

Salts yield what types of ions? _____

## 794

A solution that conducts electricity is called _____.

an electrolyte

## 795

A solution that does not conduct electricity is called _____.

a nonelectrolyte

## 796

Which of the following are anions? _____

Cations? _____

$$Na^+ \qquad Cl^- \qquad Mg^{2+}$$
$$SO_4^{2-} \qquad K^+ \qquad NO_3^-$$

anions: $Cl^-$, $SO_4^{2-}$, $NO_3^-$; cations: $Na^+$, $Mg^{2+}$, $K^+$

## 797

Saliva has a pH between 5.5 and 6.9. What type of liquid is saliva?

_____

a weak acid

## 798

Blood has a pH range of 7.35–7.45. What type of liquid is blood?

_____

a weak base

## 799

An acid of pH 2.75 is how many times as strong as one of pH 3.75?

_____

10

**buffer**

A solution that maintains a constant pH upon the addition of either acid or base is called a(n) _____.

**sugar; water**

In a solution of sugar in water, the solute is _____ and the solvent is _____.

**yes; no**

802

Do solutions pass through membranes? _____ Do colloids? _____

**100**

803

A 10% glucose solution will contain 10 g of glucose in _____ mL of solution.

**1 liter**

804

A 1M KCl solution will contain 1 mole of KCl in how much solution? _____

**0.5M NaCl**

Which solution will have a higher osmolarity, 0.5M NaCl or 0.5M glucose? _____

**806**

When two atoms share electrons, they are held together by a _____ bond.

covalent

**807**

Diffusion is the flow of _____ (solute/solvent).

solute

**808**

Osmosis is the flow of _____ .

water

**809**

Is osmosis active or passive transport? _____

passive

**810**

Is diffusion a form of active transport? _____

no

**811**

A hypertonic solution has a concentration _____ (higher than/lower than/the same as) that of another solution.

higher than

**812**

An isotonic solution has a concentration _____ (higher than/lower than/the same as) that of another solution.

the same as

lower than

**813**

A hypotonic solution has a concentration _____ (higher than/lower than/the same as) that of another solution.

semipermeable

**814**

A selective membrane is _____ (permeable/semipermeable/nonpermeable).

opposite directions

**815**

Do osmosis and diffusion normally take place in the same direction or in opposite directions? _____

K$^+$; Na$^+$

**816**

Which ion, Na$^+$ or K$^+$, has a higher intracellular concentration in animal cells? _____ Extracellular concentration? _____

ATP

**817**

What high-energy compound frequently supplies the energy for active transport? _____

Na$^+$out; K$^+$ in

**818**

During active transport by the Na/K pump, which ions are transported out of the cell? _____ Into the cell? _____

## 819

What are the relative numbers of $Na^+$ and $K^+$ ions transported out of and into a cell? _____

3 $Na^+$ out; 2 $K^+$ in

## 820

Because of active transport, the interior of the cell has a relative _____ (+/−) charge.

−

## 821

When a nerve cell is resting, are its sodium channels normally open or closed? _____ Its potassium channels? _____

both closed

## 822

When a stimulus affects a resting nerve cell, which channels open?

_____

sodium

## 823

When a stimulus affects a resting nerve cell, which ions move, and do they enter or leave the cell? _____

$Na^+$ enter

## 824

When a typical resting nerve cell has a stimulus applied, it becomes _____ (polarized/depolarized).

depolarized

Na$^+$; K$^+$

After a nerve cell has been depolarized, the _____ channels close and the _____ channels open.

repolarization

This latter change is called _____.

K$^+$

**827**

During the undershoot, which channels are slow to close?

_____

active transport

**828**

After the undershoot, which process returns the cell to its resting state?

_____

Ca$^{2+}$

**829**

When a new impulse reaches the presynaptic membrane of an axon, it depolarizes that membrane, causing an influx of _____ ions.

they cause vesicles to fuse to presynaptic membrane

The calcium ions have what effect on the vesicles near the presynaptic membrane? _____

## 831

When the vesicles fuse with the presynaptic membrane, they release
_____, which travel across the synaptic cleft to the
_____ membrane.

neurotransmitters;
postsynaptic

## 832

What effect do the neurotransmitters have on the postsynaptic
membrane? _____

_____

_____

they cause ion channels
to open, which in turn
brings about the depolar-
ization of the postsynaptic
membrane and thus the
transmission of the
impulse along the next
axon

## 833

Diagram the structure of the hydrocarbon compound containing the
following arrangement of carbon atoms, and indicate all the hydrogen
atoms.

## 834

Indicate the hydrogen atoms attached to the following arrangement of
carbon atoms.

$$C \equiv C - C$$

$CH_3CH(CH_3)CH_2CH_2CH_3$

## 835

What is the simplified structure of this compound?

OH
—CH₃

## 836

What is the simplified structure of this compound?

**polar**

## 837

Is the water molecule polar or nonpolar? _____

**hydrogen**

## 838

What types of bonds are present *between* water molecules? _____

## 839

What type of bond is involved in maintaining the coil shape of a protein? _____

hydrogen

## 840

When DNA replicates, it "unzips," breaking what type of bonds? _____

hydrogen

## 841

Are hydrogen bonds strong or weak? _____

weak

## 842

In a protein, hydrogen bonds occur between an H attached to a(n) _____ atom and an O that is part of a(n) _____ group.

N; C$=$O

## 843

What type of functional group is represented by the following?

a.  OH _____

b.  COOH _____

c.  C—C—C _____
        $\|$
        O

d.  OPO$_3$H$_2$ _____

e.  NH$_2$ _____

f.  CHO _____

a. hydroxyl
b. carboxyl
c. carbonyl
d. phosphate
e. amino
f. carbonyl

Identify the types of functional groups present in each of the compounds in problems **844–846**.

carboxyl, amino, disulfide

## 844

$$COOH \qquad COOH$$
$$NH_2CH \qquad NH_2CH$$
$$H_2C-S-S-CH_2$$

sulfhydryl

## 845

$$CH_3CH_2CH_2SH$$

a. amino and carbonyl
b. carbonyl and hydroxyl
c. hydroxl and carboxyl

## 846

a.

b.
$$
\begin{array}{c}
H \\
| \\
C=O \\
| \\
H-C-OH \\
| \\
H-C-OH \\
| \\
H-C-OH \\
| \\
H-C-OH \\
| \\
H
\end{array}
$$

c. $CH_3CH(OH)COOH$

a. _____     b. _____     c. _____

## 847

Which of the following is(are) structural isomers? _____

a. $CH_3 CH_2 CHO$    $CH_3CH_2 COOH$

b.

c.

## 848

Which of the following is a *cis* isomer? _____ A *trans* isomer? _____

a.

b.

b

a, b

a, b, c

849

Which of the following pairs are enantiomers? _____

a.

b.

c.

d.

$$CH_3CH_2-CH_2-\overset{\overset{\displaystyle CH_3}{|}}{C}=\overset{\overset{\displaystyle H}{|}}{C}-CH_3$$

$$CH_3CH_2-CH_2-\overset{\overset{\displaystyle CH_3}{|}}{C}=\underset{\underset{\displaystyle H}{|}}{C}-CH_3$$

850

Asymmetric carbons have how many different groups attached to them?

_____

4

851

How many asymmetric carbons are present in the following compound?

_____

3

C<sub>12</sub>H<sub>22</sub>O<sub>11</sub>

a. *ose*
b. *ase*

ribose; glucose

with one less oxygen

two monosaccharides

## 852

Which of the following compounds is a carbohydrate?

___ $C_6H_6$       ___ $C_{12}H_{22}O_{11}$
___ $C_6H_7ON$     ___ $C_2H_6O$

## 853

a.    Monosaccharides have names ending in _____.

b.    Most enzymes have names ending in _____.

## 854

An example of a pentose is _____. An example of a hexose is
_____.

## 855

The prefix *deoxy* means _____.

## 856

Hydrolysis of a disaccharide yields _____.

## 857

The bond holding the two halves of a disaccharide together is called a(n) _____ bond.

glycosidic

## 858

Starch is a _____ (monosaccharide/disaccharide/polysaccharide).

polysaccharide

## 859

a. Which type of carbohydrate can be absorbed directly into the blood? _____

b. Which type of carbohydrate forms a colloidal dispersion in water? _____

a. monosaccharides
b. polysaccharides

## 860

Which are unsaturated, fats or oils? _____

oils

## 861

The products of fat hydrolysis are _____ and _____.

fatty acids; glycerol

fatty acids; glycerol; phosphoric acid; a nitrogen compound

**862**

A phospholipid, on hydrolysis, yields _____, _____, _____, and _____.

cholesterol or bile salts or sex hormones

**863**

An example of a steroid is _____.

C, H, O, N

**864**

Proteins always contain which elements? _____

amino acids

**865**

The hydrolysis of protein yields _____.

many; peptide

**866**

A polypeptide contains _____ amino acids held together by _____ bonds.

867

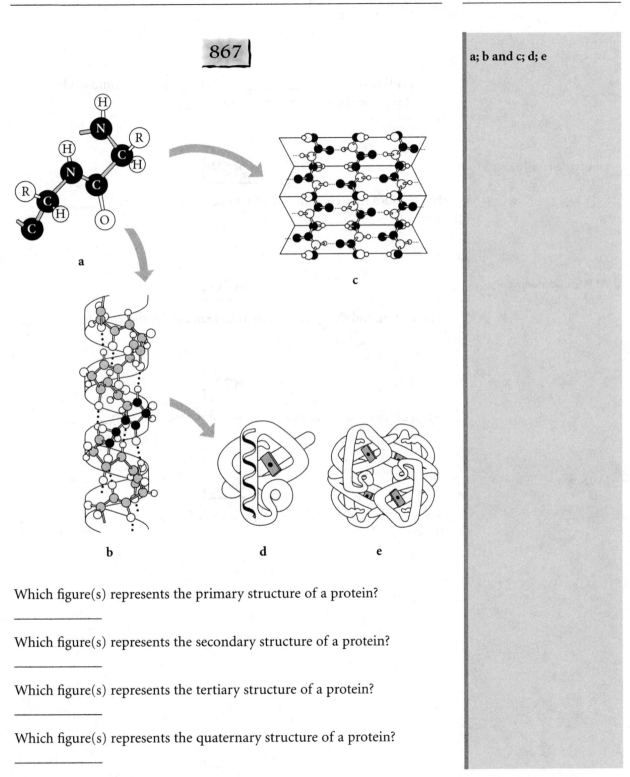

a

c

b

d

e

Which figure(s) represents the primary structure of a protein?

_____

Which figure(s) represents the secondary structure of a protein?

_____

Which figure(s) represents the tertiary structure of a protein?

_____

Which figure(s) represents the quaternary structure of a protein?

_____

3; peptide

A tripeptide consists of _____ (how many?) amino acids linked together by what type of bond? _____

ribose; deoxyribose

Nucleotides may contain one of two pentoses. They are _____ and _____.

ATP (or adenosine triphosphate)

The compound that provides chemical energy for the body is _____.

2

**871**

ATP contains how many high-energy phosphate bonds? _____

hydrogen

**872**

The opposite sides of the DNA coils are held together by _____ bonds.

## 873

a. In DNA, the partner of A is _____.

b. As RNA is synthesized, the presence of A in DNA will specify the incorporation of _____ in the RNA.

c. As RNA is synthesized, the presence of C in DNA will specify the incorporation of _____ in the RNA.

d. In DNA, the partner of C is _____.

a. T
b. U
c. G
d. G

## 874

Complete the following diagram indicating partners and bonds in the formation of mRNA.

```
DNA    ┬──┬──┬──┬──┬──┬─
       C  A  G  T  T  C


mRNA   ─────────────────
```

```
┬───┬───┬───┬───┬───┬
C   A   G   T   T   C
⋮   ⋮   ⋮   ⋮   ⋮   ⋮
G   U   C   A   A   G
┴───┴───┴───┴───┴───┴
```

## 875

Which amino acids are designated by the mRNA in problem 874? (See table on page 213.) _____

val, lys

mutation

**876**

If a genetic code is miscopied, such a change is called a(n) _____.

no

**877**

Are all mutations harmful? _____

specific

**878**

Enzymes are _____ (specific/nonspecific).

coenzyme

**879**

The nonprotein part of an enzyme is called the _____.

a. competitive
    inhibitor
b. noncompetitive
    inhibitor

**880**

a.  A substance that "fits" into the active site of an enzyme and prevents that enzyme from functioning is called a(n) _____.

b.  A substance that alters the shape of the "active" site of an enzyme by reacting with that enzyme at a spot other than the active site is called a(n) _____.

**881**

What is a transferase?

_____

an enzyme involved in the transfer of a substance from a donor to an acceptor

**882**

What is an aminotransferase?

_____

an enzyme involved in the transfer of an amino group from a donor to an acceptor

**883**

What is an N-methyltransferase?

_____

an enzyme involved in the transfer of a methyl group from a donor to a nitrogen in an acceptor

**884**

Give the structure for
a.  3,3-dimethylbutylamine

a.

$$CH_3-\overset{\overset{\displaystyle CH_3}{|}}{\underset{\underset{\displaystyle CH_3}{|}}{C}}-CH_2-CH_2-NH$$

b.  N,N-dimethylpropylamine

b.

$$CH_3-CH_2-CH_2-\overset{}{\underset{\underset{\displaystyle CH_3}{|}}{N}}-CH_3$$

**succinate; FAD**

### 885

Which of the two acids in the diagram is oxidized? _____

Which substance is reduced? _____

**pyruvate**

### 886

Glycolysis involves the conversion of glucose to _____ .

**cytoplasmic fluid**

### 887

Glycolysis takes place in which part of the cell? _____

**anaerobic**

### 888

Is glycolysis an aerobic or anaerobic sequence? _____

### 889

Does glycolysis produce ATP? _____

yes

### 890

The NADH produced during glycolysis is transferred to which sequence?
_____

electron transport chain

### 891

Pyruvate is converted into _____ in which part of the
cell? _____

acetyl CoA;
mitochondrion

### 892

Acetyl CoA enters the _____ cycle, which takes place in which
part of the cell? _____

citric acid; mitochondrion

### 893

Is the citric acid cycle an aerobic or anaerobic sequence? _____

anaerobic

### 894

Is ATP produced during the citric acid cycle? _____

yes

electron transport chain

The NADH and FADH$_2$ produced during the citric acid cycle are transferred to which sequence? _____

mitochondrion

The electron transport chain occurs in which part of the cell? _____

aerobic

Is the electron transport chain an aerobic or anaerobic sequence?

_____

glycolysis, citric acid cycle, electron transport chain

In which parts of the cellular respiration sequence is ATP produced?

_____

light

899

What is the source of energy for photosynthesis? _____

carbon dioxide, water

900

Which reactants are involved in photosynthesis? _____

water

From which compound is oxygen produced? _____

## 902

Which photosystem can be cyclic? _____ Which is noncyclic?
_____

I; II

## 903

Which compounds are produced in the light reactions? _____

ATP, NADPH, $O_2$

## 904

ATP is produced from which compound? _____

ADP

## 905

NADP is produced from which compound? _____

$NADP^+$

## 906

Which compound enters the Calvin cycle? _____

carbon dioxide

## 907

Which compound is produced by the Calvin cycle? _____

glyceraldehyde-3-P

## 908

Which high-energy compounds are produced by the Calvin cycle?
_____

ADP, $NADP^+$

no; yes

### 909

Is oxygen produced in light reactions photosystem I? _____ In light reactions photosystem II? _____

higher; lower

### 910

Gases flow by diffusion from an area of _____ (higher/lower) partial pressure to an area of _____ (higher/lower) partial pressure.

oxygen; carbon dioxide

### 911

In the lungs, arterial blood picks up _____ gas and gives off _____ gas.

hemoglobin

### 912

What substance in a red blood cell transports oxygen? _____

oxyhemoglobin

### 913

When hemoglobin reacts with oxygen, what compound is formed?

_____

yes

### 914

Is hemoglobin recycled? _____

## 915

Carbon dioxide is a waste product produced in the tissues. It flows into the red blood cells, where it reacts with water to form _____.

**carbonic acid**

## 916

Carbonic acid ionizes to form the _____ ion and the _____ ion.

**bicarbonate; hydrogen**

## 917

The flow of bicarbonate ions from the red blood cells into the blood plasma and the opposite flow of chloride ions from the plasma into the red blood cells are called the _____.

**chloride shift**

## 918

Where in the body does the reverse chloride shift take place?

_____

**in the arterial blood in the alveoli**

## 919

Carbon dioxide is carried in the blood in two forms, the _____ ion and the compound _____.

**bicarbonate; carbamino-hemoglobin**

# Conclusion

Congratulations! You have completed the program. If you have worked conscientiously, you should now be able to

- recognize the elements present in various biological compounds

- understand the term *pH* as applied to fluids and cells

- recognize what is meant by the terms *acids*, *bases*, and *salts* as they occur in plant and animal tissues

- know what electrolytes are so that their functions in the life of the cell can be more clearly understood

- be able to draw Lewis dot structures

- recognize organic functional groups

- recognize types of isomers

- understand oxidation and reduction as they occur in the metabolic processes of plants and animals

- differentiate among carbohydrates, lipids, and proteins, and recognize each from its unique functional groups

- understand how enzymes function

- know what nucleic acids are and how they are held together

- understand diffusion and osmosis

- distinguish between passive and active transport

- understand osmotic pressure

- understand transmission of nerve impulses, including depolarization and repolarization

- ❧ understand how neurotransmitters work

- ❧ know how DNA replicates

- ❧ know how mRNA is formed and how it regulates protein synthesis

- ❧ understand the function of methyltransferases

- ❧ recognize the structure of N-methyl compounds

- ❧ understand methylation and demethylation reactions

- ❧ follow the flow of oxygen from the lungs to the cells and the flow of carbon dioxide from the cells to the lungs based on gas partial pressure gradients

# Appendix

# Figure A
# Cellular Respiration

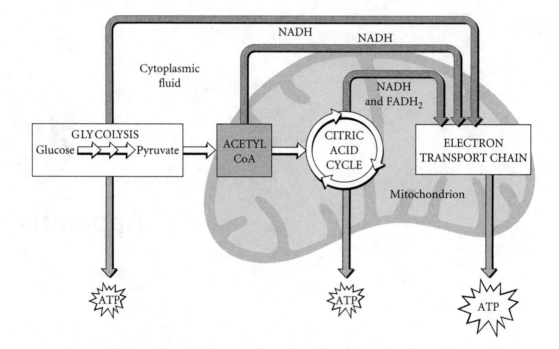

# Figure B
# Glycolysis

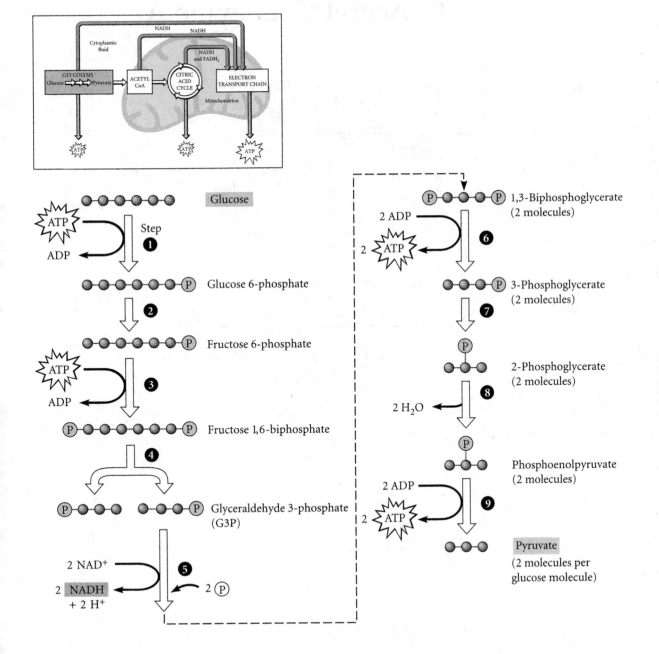

# Figure C
# Conversion of Pyruvate
# to Acetyl Coenzyme A

# Figure D
# The Citric Acid Cycle

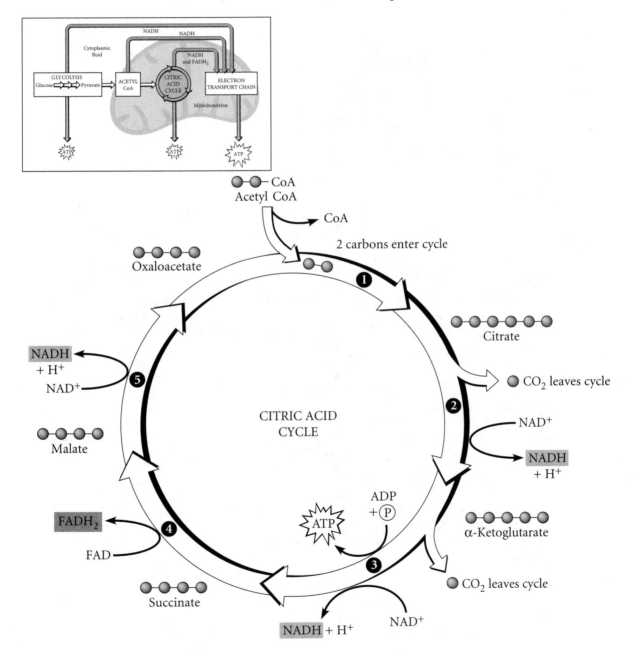

# Figure E
# Electron Transport Chain

# Index